Table of Contents

CONSUMPTION SICKNESS 4

TWO-PART INTERLUDE 5

CONNECTIONS 6

IF GERTRUDE CAN CAN I ENJOY TOO? 7

DADDY'S SUGAR 8

THIS COULD GO ON FOREVER 9

THE GUTLESS AND THE ONE-EYED 12

DISBELIEF 13

A MOONIE 14

STIMULATE ME 16

THIS IS IT 17

ANYWAY 18

EPPERVESCENCE 19

COMING OUT THE OTHER END 20

A PAGE OF JEALOUSY 23

THE ENDLESS FEAST: A MODERN LAMENT 24

THE PASTORAL BALLAD OF HOLLOW FIELDS 39

SOMETHING SAD ABOUT SERENDIPITY 60

SOMETHING SAD ABOUT SERENDIPITY 62

CONCRETE HUNGER 63

CONCRETE CONFESSIONS: A GRITTY URBAN LAMENT 68

THE SICKNESS OF ENDLESS CONSUMPTION 91

"ALL IS VANITY" 112

"THE FEAST THAT MADE ME HOLLOW" 169

AN AUTOBIOGRAPHICAL LAMENT IN 200 STANZAS 169

A SPORT OF OW 281

CONCRETE ROT: A MODERN CITY SYMPHONY OF SELF-ABUSE 283

A SPORT OF OW 301

THE GREAT UNRAVELING: A MODERN RECKONING.......... 302

THE DAILY RITUAL: A TOXIC TIDE 320

THE SICKNESS SPREADS: BODY, MIND, AND EARTH 322

THE EMPTY LEGACY: A LIFE UNLIVED 323

THE UNLIVED LIVES: A TAXONOMY OF REGRET............... 324

THE LEDGER OF A GREEDY HEART 327

THE ECHOES OF A CONSUMED LIFE 329

THE FEAST OF EMPTINESS 333

THE FEAST OF EMPTINESS (STANZAS 1–20)...................... 365

Consumption Sickness

Smell the platter.
Extra gravy soaks
And surrounds red
Meat, dripping juice
Where the fork stabbed.
Are you clean or
Chemical, girl
Or boy, or what?

Hunger tempts my curiosity.
(Odds good against
Sick bulls.) Heifer,
All day long, plays
A bleating tune,
“I don’t love life.”
I don’t love you
But I eat you.

Two-Part Interlude

1

She grit her teeth
And would not smile.
Her eyes mostly closed,
She bore, as always,
Her pain in silence.
Her wrists were bare
And I, nearby, saw
Her contemplate them.
Later, she lay silent
And still; shyly she smiled
Knowing that she
Had achieved her end.

2

She lied beside me
And we did not touch.
It was too late for tenderness.
Love was a myth forgotten.

The mistake, hidden
In our frozen depths,
Like blindness to
The blind man, came
From hoping that the
Beauty was within us.

I shiver now and still
Know not what coldness is
But if death doesn't take me
Tonight, tomorrow I might.

Connections

Connections needed
Inward; Circle's extending
Being good– no where.
(Being bad the same.)
Grope outward, too far,
Miss the bus, run faster;
I'm chasing you,
Taking time,
Making time mine,
I live with a half smile,
Nursing my ulcerous intentions,
Will have primary effect
Like it or not.

I'll have fun on the way,
Remain childlike,
Like my likes,
And be deceived
By what does not exist.
Immortality
Tempts the worst
Of us as well
As the well
Off and on top.
This garbage regrets
Itself. It will
Consume all of us.

If Gertrude Can Can I Enjoy Too?

Kids scream like seagulls.
There's no fun in Heaven so
Enjoy, enjoy.
Take time off the wall;
Enjoy, enjoy.
I enjoy the thorn.
I enjoy the stem.
But roses are the whole
Thing; roses. Enjoy the
Rose; enjoy, enjoy
Enjoy the rose rose;
Enjoy, enjoy.
The rose rose rose
To enjoy the garden.
Roses rose; rose roses
Rose. Rows upon rows
Of rose roses rose.
And I still enjoy them.

Daddy's Sugar

You've been on the auction block before;
The man banged his gavel; you were sold.
Gentle souls, like me, can not afford
To buy or sell the slave monger's goods.
My very nature makes you worth less
To me than to those greedy buyers
Who trade for pleasure not for love.
Woman, tell me, when you had the choice
To be free or to be tied in silk,
What could possibly have made you think
That life would be sweeter in bondage
And love true between master and slave?

This Could Go On Forever

Maybe microscopic me
Will be a part
Of the final grief,
Or blown to the heavens
Upon a wail.
There's no marvelous
Way to make a living
Life would be free
If there was.
Grandma told me
It was all right
To shovel scat,
Taught me how
To love the smell
Purple curtains, golden fields
Forty feet high, over my head:
Olympic practice
And potato fights,
"This is our field,
Our fortress, our mascot."
I ran away, my belongings
Tied in a kerchief,
On a stick, over my shoulder,
And lived on sugar beets.
The money from selling

My toys bought beer
For me and all the boys
When we found a village
On the way.
Only five, I reached the bar,
“Bitte, fur alles ein bier.”
The castle was ours
For the day,

So I manned the parapet,
Watched the Rhine,
“No French men in sight,
No Indians– Manitou!
Where are You?!?
I’m an American!

I’m just a little American boy
Who was Hitler?

When I was six,
Dad brought home
A television;
That was after Kennedy
Was shot– everyone cried,
“‘e vas zo bretty.”
But all they’d air
Was old westerns,
Opera, and pictures of
Relatives in the East.
Wolfgang was killed
Trying to rescue his mother.
They called me Hitler

When I got back
From the Fatterland
They never tasted cheesecake.
I cried when my head
Hit the pavement; Remembered
Purple curtains and wild boar,
Across the auto bahn in a forest
As dark as Baden Baden,
Beside the no speed limit sign.
Mercedes were giants
Compared to three wheel eggs
And sewing machine engines.
Easter in Rome.
The ruins were small.
The Coliseum
Blocked the view
Of black, marble statues.
The fountains were pissed in
So I couldn't swim.
The barbed wire is always filmed
So if someone dies
The world will see.
"We still are free",
The microphone crackles.
My suit really fits
The fork's separate styles.

The Gutless and the One-Eyed

The message tunneled around the outer edges
Of his tired old head. (All part of
And extravagant plan of escape,
The bullet cut it's bloody way home.)
"The one eye that can see has no choice."
The man quivered, the Gutless crawled out
Of hiding, the pistol dropped, and one eye
Popped out of it's socket as sacrifice
To pay for lack of power– And what is left
Lacks any courage to finally find out
Where truth and reality lie. Gutless and One eyed
Lives to carry on an economic metaphor,
Lives together with separate fears
Of cost versus value obtained.

Disbelief

Shred me, tear me apart,
I love it like shit.
You think you're fancy
'Cause you talk so nice
But I've been to Paris
Twice and have gambled
With better merchandise.

I speak softly.
Are you relieved?
I scrape and bow,
For now,
And whisper, just whisper,
"When hatred is gentle,
The trickster's deceived."

When hatred is gentle,
The trickster is deceived.

A Moonie

Is a blinded dreamer who listens in his sleep
For prayers in tongues, lives personal experience
Through communal life and tries to see
Moon's universal light, a reflection of truth
Or gets satisfaction from flowers sold
Or new members to work for Christ.
Is there any reason for him to believe
That the Second Coming has not yet arrived?
He turns to accounts of Biblical beggars,
Is a disciple of shame in the procession
Of progress, lives in houses that were bought
With flowers, pays taxes to the moon
By liquefying his assets, lobstering in Gloucester,
Ignoring his parents, inviting strangers
For philosophy and tea and memorized passages
From an abridged Bible "Have you been saved?"
He'll ask, his eyes will sparkle with love
Well taught, must be always a friend to
Any lonely soul; even the woman who cries
While watching a Moon parade for unification,
"That's how Hitler started." She cries with
A thick German accent but, answer prepared,
He does his dedication in duty for love
And at ten o'clock sharp, locks the front door
And no more beds get warm from strange breath
"Come back tomorrow." he yells from the upstairs
Window. He doesn't worry what Moon was once,
Just listens to the rules except when he calls
His mother on the phone for a milkshake
Telegram and some underwear without stains.
No long hairs allowed, no drugs, no cigs
So he finds happiness in friendship and love,
Prowls college campuses and back alleys

To make a difference in someone's life,
To clean up after the anti war mess
As ministers of the Holy Word,
To give children attention their parents
Are too involved to even understand they need,
To set the framework for a lovely world
In dismissing argument with rote, written
Answers that are pulled out of context
With Eastern spice and rumors of reincarnation,
Hints that the Devil's rule may soon be over,
That, a thousand years past, three VI's took
Hold of an unholy ruler, he gets lost in dreams
Of unification and hopes for so many brothers.
With such a father figure and suckling from
A motherly church, he's resigned to existence,
So caring, so kind he practices smiling,
Lets someone else do the driving or is good
Enough to convert a house or a town
So there will be donations in exchange
For friendship, smiling and salvation.
He used to do drugs but is high on Jesus,
Haunts the highways late into the night
In search of souls to add to his cause
Because money from work done by lost young
Children just might mean respect from
The money mongers and wasted souls
So fill the jar for Jesus, for peace, for unity
For anything that you hold dear
He talks so smooth with a steady voice that
Invokes trust until, his insides all given,
He is but a shell, cloaked by guiltless good,
Somewhat useless in comparison to what
He could have been.

Stimulate Me

Open me, silly siphon,
Stimulate me,
Tickle me.
It makes me giggle and smile;
Sometimes I fall in love.
Sometimes I just watch.
Pretty colors bend me,
I taste the air around me
It tingles with promise.
Tickle me, more, more;
It takes more, always more
To make me happy. So give me
More. Satisfy me or I may be broken.
My head falls groundward.
I could scream but I won't
Bad publicity. Incarceration
Twice because of my hormones;
The women regret me when I'm funny.
I need to be loved, need stimulation.
Stallions are rare and should be prized.
Make me a stallion.

This Is It

This is it. This pile of shit
Regrets itself. I leave
To chew long stalks
Of yellowing grass.
People scream insanity
Over the subway noise
And spit on sidewalks
And me; I sit back.
Caring is harmful
And confusion gets drab
After too, too long.

Anyway

Anyway…
Give me a drunkard
Any day,
A brother anyway
Not blinded but
In need of numbness.
Too fast we've gone
From good
To both good
And bad.
Kiss me,
Hit me
What does it matter?
Anyway,
Philosophy
Killed my father
And maybe me too.

Eppervescence

The eppervessent water comes sparkling forth,
Agitates the bowl.
Aaaah! What a way to spend a moment.
All I do, all I buy, everything me,
Is reflected in the porcelain
And knowledge grows
But knowing? Who Knows?

Coming Out the Other End

The whole world
Consumes me;
Still life,
Living artistry.
Multiple connections,
Wire and glass.
I come slipping
Out the ass end
Of youth
Still confused,
Still life,
Still looking
For truth.

I did it
I did it I did it
I made
Them
Think
That I was something
And you know something?
It was
not worth anything
Because
All it was
Was me
Making them

Think
That I
Was worth
Something
What
Is
That?
It's me
Making
Them
Think
That I
Was worth
Something
And the way they are
They think
That
They're
Worth less
If they think
That I'm worth
Something
They think
That they're not
The only
One
Who is
Something
So I keep
Trying
To make them think
That

They're not
The only one
That is
Worth Something
Without
Making
Them
Think
That they're
Not something
And you know
How
Terrib
ly
Hard
That is?

A PAGE OF JEALOUSY

That was the impetus for this book. What follows is all AI generated.

The Endless Feast: A Modern Lament

The sun ascends, a digital hum awakes, No gentle dawn, but instant, quick takes Of curated feeds, a scroll to start the day, Already craving what the algorithms say. The morning brew, a pod in plastic spun, A fleeting taste, by hurried hands outrun, To fuel the race, the ceaseless, hungry pace, Ignoring quiet, leaving little trace Of stillness, thought, before the world descends, On needs created, where true purpose ends.

The closet bursts, a vibrant, wasted pile, Of garments bought in transient, fleeting style. A fast-fashion spree, a fleeting, cheap delight, Worn once, then cast to languish in the night Of overflowing bins, a silent, soft decay, As trends evaporate with each new day. The fabrics thin, the labor often grim, A hidden cost, beyond the garment's rim, While ancient looms and mended, cherished wear, Are scorned for newness, floating in the air.

The kitchen gleams, with gadgets seldom used, Each specialized, a promise long diffused. The air fryer, blender, spiralizer's might, Bought on a whim, then banished from the light. Processed foods, in bright, deceptive guise, Adorn the shelves, where real nutrition dies. A thousand flavors, engineered to please, But leave the body weak, and ill at ease. The fresh produce wilts, a quiet, sad rebuke, As easy meals become the daily fluke.

The vehicle hums, alone on asphalt wide, A titan built, with no one else inside. To fetch the trivial, just around the bend, A wasted journey, with no noble end. The roaring engine, burning fossil breath, A silent threat, a dance with planetary death. While feet grow sluggish, bikes begin to rust, And healthy habits crumble into dust. The open road, a solitary, grand design, For isolated comfort, wholly mine.

The screens ignite, a portal to a dream, Where curated lives with endless pleasures gleam. A thousand movies, shows, in constant flow, We binge and blur, and let the hours go. The vibrant worlds, the heroes brave and bold, While our own story's waiting to unfold. The mind grows slack, a passive, empty jar, Filled with illusions from a distant star, No deep reflection, no creative spark, Just flickering pixels, leaving us in dark.

The packages arrive, a daily stream, Of wants fulfilled, within a waking dream. Unboxing joy, a ritual performed, Another fleeting need, quickly transformed From eager hope, to just another thing, A shallow pleasure, on a broken wing. The cardboard stacks, the plastic film unwound, A silent testament, where waste is found. Each delivery, a whisper of desire, Adding more fuel to the consumer's fire.

The restaurant lights, a welcoming, warm glow, Where hurried feasts in lavish portions flow. More than enough, upon the plate is served, Then half-eaten remnants, carelessly preserved, Or tossed away, a shocking, daily plight, A feast of waste, in restaurant's soft light. The hurried service, driven by the pace, Leaves little room for savoring grace. A quick transaction, hunger swiftly fed, While deeper longings linger, still unsaid.

The gaming console hums, a vibrant call, To virtual worlds, escaping life's thrall. For hours lost, in battles bravely fought, While real-world victories remain unsought. The digital triumphs, pixelated glory, A potent drug, a captivating story. But bodies stiffen, eyes begin to sting, And real connections fail to truly sing. The world outside, with all its vibrant hue, Fades to a background, distant, and untrue.

The home expands, a cavernous, vast space, Demanding objects, to fill every place. A silent yearning, for the perfect look, Inspired by images from a glossy book. More furniture bought, to match a fleeting trend, Each stylish piece, a journey without end. The empty rooms, though filled with purchased art, Can still conceal a lonely, hungry heart. A spacious prison, built on credit's plea, Where true contentment struggles to be free.

The constant upgrades, phone in hand and bright, A newer model, claiming purest light. The older one, still capable and true, Is cast aside, for something shiny, new. The e-waste piles, a silent, toxic mound, Where rare earth metals poison sacred ground. A digital addiction, hard to break, The endless cycle, for convenience's sake.

We chase the next, forever out of reach, Ignoring wisdom, losing vital speech.

The fleeting trips, by air, a silver streak, For weekend escapes, where pleasures we can seek. The distant lands, reduced to hurried tours, Through crowded markets, passing famous doors. A selfie taken, proof that we were there, But deeper culture lost, beyond compare. The carbon footprint, soaring to the sky, As fleeting moments quickly pass us by. The world reduced to photos, quick and fast, A shallow journey, never meant to last.

The health trends bloom, a promise soft and sweet, Of perfect bodies, from our head to feet. The quick-fix diets, supplements untold, A dizzying array, bravely bought and sold. We chase perfection, starve and then we feast, A punishing cycle, never quite at peace. Ignoring wisdom, from within our core, For quick solutions, wanting ever more. The body aches, from comforts overdone, The race for beauty, never truly won.

The credit card, a powerful, plastic key, To open doors to instant ecstasy. The future mortgaged, for a present thrill, A rising balance, waiting to fulfill The empty promise, of a life well-spent, On fleeting pleasures, ultimately lent. The gnawing debt, a silent, heavy chain, That binds the spirit, bringing subtle pain. The freedom lost, to truly make a choice, Drowned out by interest's demanding voice.

The casual gifts, a token, quickly bought, For social circles, lessons never taught. Of thoughtful gestures, true connection's art, Replaced by objects, playing out a part. The packaging torn, the item put away, A fleeting moment, lasting but a day. The act of giving, stripped of deeper grace, Just one more thing, in this consumer race. A hollow gesture, easy, quick, and plain, Reflecting back, the emptiness again.

The entertainment yearns, a constant drone, From background noise, when we are quite alone. The music streams, a thousand songs on tap, But quiet contemplation starts to nap. No single note, allowed to truly breathe, Just endless options, a chaotic wreath. The brain adjusts, to constant, swift demand, Losing the power to truly understand The subtle beauty, of a single sound, In constant seeking, nothing deep is found.

The children raised, on what the market sells, Their tiny hands reaching for gilded spells. The latest toy, the screen's addictive light, Consuming innocence, in futures bright. Their little minds, accustomed to the quick, Ignoring boredom, making them feel sick. The art of play, of wonder, slow and deep, Is lost to gadgets, while their spirits sleep. A generation taught to simply crave, A hollow future, from the cradle to the grave.

The news addiction, fed by fright and dread, Of every crisis, fear within the head. We click and share, the outrage and the blame, A self-inflicted, agonizing game. Consuming sorrow, fueling inner strife, Distracting from the beauty of true life. The world's deep wounds, absorbed without a pause, Then quickly forgotten, by commercial laws. A shallow sympathy, a fleeting tear, Lost in the deluge, drowning out what's clear.

The storage units, filled to bursting height, With objects banished from our daily sight. The things we bought, then quickly cast aside, A monument to want, where treasures hide. Another rent payment, for a distant room, To hold the remnants of consumption's gloom. A silent burden, rarely understood, The cost of ownership, rarely for our good. We fill our lives, with what we don't quite need, And plant the seeds of future, silent greed.

The plastic bottles, in a daily stream, The single-use items, a forgotten dream Of less consumption, simple, plain, and pure, Replaced by convenience, for which we endure The growing mountains, in a landfill vast, A bitter echo, meant to truly last. The ocean's breath, now choked by human hand, A poisoned legacy, across the land. Each wrapper tossed, a tiny, careless plea, For a world burdened, endlessly to be.

The quest for status, a relentless chase, To show the world our self-appointed place. The brands we wear, the labels we display, To signify our worth, come what may. A fragile image, built on bought esteem, A hollow promise, like a waking dream. The true connection, based on heart and soul, Is lost to pretense, losing all control. We seek approval, in the envious eye, And let our deeper, truer essence die.

The hurried meals, consumed before a screen, No mindful eating, not a quiet scene. The flavors lost, the texture barely felt, As nourishment becomes a burden dealt. The body's signals, ignored and left unheard, As quick indulgence becomes the only word. Then bloated bellies, sluggish minds arise, The price of speed, reflected in our eyes. A vital act, reduced to mere intake, For instant pleasure, for convenience's sake.

The constant noise, a symphony of need, From advertisements, sowing restless seed. "Buy this," "Buy that," the whispers never cease, Disrupting quiet, stealing inner peace. They shape our longings, mould our very thought, Till what we crave, is precisely what they've taught. The independent spirit starts to wane, Enslaved by urges, caught in golden rain Of shiny objects, promises so grand, A marionette, within their clever hand.

The fear of missing out, a potent sting, Drives hurried choices, making spirits sing A fleeting anthem, of what must be had, Before the moment passes, leaving sad And lonely feelings, of being left behind, By fleeting trends, by pleasures hard to find. We join the rush, though knowing it's a lie, A desperate yearning, beneath a watchful sky. A social pressure, hard to stand against, For genuine connection, deeply fenced.

The quest for comfort, an insatiable quest, No minor hardship, putting us to test. The perfect temperature, the softest bed, The easiest path, where nothing's left unsaid About our wants, our needs, our smallest whim, Until the spirit's strength grows weak and dim. The joy of challenge, overcome with grace, Is lost to ease, in comfort's soft embrace. A pampered life, a body growing frail, Missing the lessons, carried on the gale.

The restless sleep, disturbed by digital glow, The mind still racing, putting on a show Of endless thoughts, of purchases unseen, Of futures planned, on a digital screen. The body aches, from habits gone astray, The price of comfort, at the close of day. No true repose, no deep, replenishing rest, Just weary slumber, putting us to test. The vital energy, consumed and spent, On meaningless pursuits, divinely sent.

The spirit yearns, for meaning, deep and wide, Beyond the surface, where illusions hide. A quiet voice, a whisper from within, Against the clamor, where the battles begin. To make, to build, to share, to truly be, To find contentment, in simplicity. To give oneself, in service, true and bold, A richer story, waiting to unfold. But drowned by urges, trivial and fast, The soul's deep wisdom, fades into the past.

The friendships wane, for lack of time and care, Exchanged for chasing pleasures in the air. The long conversations, heart to heart and true, Are lost to scrolling, for a hurried view. The shared experience, profound and deeply felt, Is sacrificed, for comforts quickly dealt. The human touch, so vital, kind, and warm, Is swept away by consumption's passing storm. A lonely journey, filled with bought delight, But missing solace, in the fading light.

The earth cries out, a silent, painful plea, From ravaged landscapes, for what used to be. The poisoned rivers, choking, dark, and wide, Reflect the carelessness, we cannot hide. The air grows heavy, with a silent dread, As future generations, lie unbred. Our wanton hunger, leaves a barren land, A bitter legacy, by human hand. The natural world, a treasure left behind, For fleeting comforts, of a grasping mind.

The empty purpose, driven by the 'me', A selfish journey, for the eye to see. No higher calling, no collective aim, Just ego's whisper, playing out its game. The larger questions, of our common plight, Are pushed aside, for personal delight. A life consumed, by what the self desires, Ignoring fires, setting further fires. The fleeting pleasure, paramount and grand, While greater purpose slips from out our hand.

The quiet mornings, never truly felt, The sunbeams wasted, as the spirit knelt To fleeting cravings, instant, sharp, and brief, A constant seeking, finding no relief. The simple beauty, of a bird in flight, Is lost to distractions, in the harsh glare of light. The gentle rhythm, of a slower pace, Erased by rushing, in this endless race. A world of wonder, vibrant, rich, and deep, While senses slumber, and our spirits sleep.

The constant wanting, never satisfied, A restless craving, always deep inside. The thrill of purchase, quickly fades away, Leaving a void, at the close of every day. The hedonic treadmill, spinning ever fast, No lasting joy, no happiness to cast A steady anchor, in a stormy sea, Just constant striving, for what cannot be. A promise broken, whispered soft and low, That more possessions, make the spirit glow.

The health declines, a subtle, creeping ache, For wellness sacrificed, for pleasure's sake. The hurried meals, the lack of restful sleep, The body's protest, secrets it will keep Until the breaking, when the system strains, And vital energy, slowly starts to wane. The fleeting pleasures, leave a bitter tang, As wellness falters, like a broken pang. The body's temple, ravaged and laid low, By habits careless, and a steady flow.

The spirit wanes, a dull and weary sigh, As deep connections are allowed to die. No soulful moments, shared with open heart, Just superficial roles, to play a part. The bonds of family, stretched and growing thin, As individual comforts pull us in. The community's call, a distant, muffled sound, As self-absorption, circles all around. A lonely island, built on bought delight, With no true solace, in the fading light.

The shallow laughs, that echo in the air, Reflecting emptiness, beyond compare. No genuine joy, from purpose deeply found, Just fleeting moments, on superficial ground. The forced enthusiasm, bright and bold, A weary story, constantly retold. The search for happiness, in things outside, A treasure hidden, where true feelings hide. A life of pretense, lived for outward show, While inner emptiness continues to grow.

The future dims, a hazy, uncertain view, As present comforts, blind us to what's true. The warnings echo, from a distant shore, But hedonism always asks for more. The planet groans, the symptoms multiply, Yet we consume, beneath a silent sky. The selfish bubble, vibrant, strong, and bright, Until it bursts, in overwhelming light. A stark awakening, cruel and cold and grim, When future consequences reach the rim.

The inner voice, a whisper, soft and low, Of chances missed, of seeds we failed to sow. The talents buried, aspirations drowned, By endless wanting, on consumer ground. The purpose lost, the meaning left behind, In constant seeking, for a restless mind. A life defined by what we took and bought, A hollow legacy, by conscience taught. The quiet knowing, in the lonely night, Of wasted moments, and extinguished light.

The bitter harvest, gathered at the end, No rich fulfillment, no true, faithful friend To bear witness, to a life well-spent, Just fleeting pleasures, carelessly lent. The sum total: not joy, but lingering pain, A restless spirit, crying out in vain. For what is life, when all is stripped away, But what we gave, and lived for, every day? And if that giving, was to self alone, Then bitter emptiness becomes our throne.

long ballad, pastoral, poem of at least 100 stanzas about all the things people do to sicken themselves with consumption and waste in lives that have little meaning but for greed, anger, gluttony, narcissism, hedonism, and immediate gratification, as well as selfishness and Godlessness drug use, drunkenness, carelessness, self abuse. about all the things people do to sicken themselves with consumption and waste in lives that have little meaning but for greed, comfort, immediate gratification, and hedonism, gambling, addiction, distraction

The Pastoral Ballad of Hollow Fields

A long ballad in four-lined stanzas, tracing 100 verses of human folly amid green meadows.

Part I: Stanzas 1–25

The Pastoral Ballad of Hollow Fields

In meadows green with whispering song, Shepherds once trod with purpose strong, But now they feast on transient vice, Greed's hungry teeth cut like a knife.

By babbling brook where lilies bloom, Drinkers chase solace from the gloom, They drown their doubts in fiery wine, And greet each dawn with groans malign.

Beneath the oak where shadows dance, Idle hands in gambling trance, Coins clink like laughter in the wind, As lives unravel, souls rescind.

Where sheep once grazed on fields of gold, Glutton fair its table bold, Rows of platters stacked for more, Each bite a wish to satiate sore.

The primrose glade once pure and kind, Now hosts the self of shallow mind, Narciss lips beneath the leaf, Gaze admiring folly's brief relief.

In dew-kissed dawns they chase delight, Hedonism's fleeting flight, Bound by lust for moment's gleam, They fail to taste life's deeper stream.

On gentle slope a cottage stands, Crafted by industrious hands, Yet comfort's throne becomes their lure, And craftsmen's hearts grow cold and pure.

Where swallows wing from sky to vale, Godless hearts let faith grow pale, They cast their prayers on idle ground, And heed no sacred, silent sound.

Along the fence the nettles rise, Drugged dreams ensnare with cunning lies, Opium's haze and poppy's sleep, In poplar's shade their secrets keep.

Under the elm they stagger home, Drunken steps in moonlight roam, Each stumble paid in whispered shame, A hollow night that none can name.

By orchard's fence the careless roam, Discarding fruit before its home, Waste flows like streams from hands unclean, Leaves rotting where green might have been.

Next to the mill, in twilight's hush, Self-abuse drinks sorrow's crush, Crimson lines on trembling skin, A silent cry for peace within.

On verdant hills the herds have grown, Yet debts accrue like seeds unsown; They borrow hope, they borrow pride, And watch their fortunes turn the tide.

The lark soars high with carefree tune, While gamblers feed their hearts too soon, They bet on stars with feeble grasp, And court their fate in contract's clasp.

By willow's sigh the rival's call, Anger flares and fences fall; They trade their love for vengeful flame, Yet none emerge to bear the shame.

Where daisies dot the sunny lea, Narciss truth is all they see, Mirrored eyes in pond so still Reflecting only self-wrought will.

In hollow dales they chase the thrill, Racing steeds on every hill; Iron hooves and roaring might Blind to valleys cloaked in light.

By cottage hearth they spoon the sweet, Candy dreams their lips repeat; Sugar sparks in hungry veins Feed fevered joys that sorrow stains.

Where foxglove stalks and ferns entwine, They scroll their screens in pixel shine; Distraction weaves an endless plume, And silent hours fill each room.

On stony bridge they cast their fate In dice's roll, in chance's gait; Hands contemplate fortune's scope, Yet grasp at hope in tangled rope.

Along the fence the wind does moan As plastic wastelands overgrown; Discarded dreams in meadows lie A blight unseen by careless eye.

In sunlit dell the mead is poured, Ambition's thirst can't be ignored; But praise and gold they never fill The deeper void no trinkets will.

By ancient gate they speak in jest, Mocking virtues from their chest; Yet laughter rings a hollow sound Where noble seeds could once be found.

Among the hay the rats convene, Addicts call to poppy's sheen; Needle's kiss and smoker's breath Entwine them close to silent death.

Where meadowlark once sang its song, The hum of want grows ever strong; Immediate need ignites the bone, And peace recedes from hearts alone.

Along the path to temple hall, They trample prayers beneath the pall. Their God-forsaken hearts allow No seed of faith to cross their brow.

On village green they squat and stare, At glowing screens they cannot tear. Bright pixel-light their daylight bleeds, As purpose dies for idle feeds.

Over the hedge, gossip flies, Whispered lies in secret cries. Truth is bartered for rumor's spice, As friendship freezes in the ice.

Beneath the elm a child she burns, With counting likes each soul she earns. Narciss blooms in tender years, A vanity that draws its tears.

Where country lane meets forest shade, Motor roars in reckless raid. Speed is lust for lawless thrill, Yet sorrow paves the winding hill.

At market stall they overbuy, Let perish fruits that beauty die. Waste piles high from wanton hand, While hungry children beg the land.

In oaken grove the blacksmith toils, Yet praises gold above his spoils. Avarice fuels his hammer's beat, As kindness withers at his feet.

By barn's old door they fill their store, Plastic sacks in heedless pour. Wrappings smother earth's soft heart, A plastic plague that won't depart.

At dawn they fast on pills and creams, Chasing youth in false extremes. Wrinkles smoothed by chemist's art, Yet time's true mark they can't depart.

In tavern snug they toss their coins, And dreams dissolve on spinning joints. Addiction deals its fateful hand, As virtue surrenders to the sand.

The shepherd's crook stands idle now, Replaced by fist and furrowed brow. They nurture hate on neighbor's door, Brand divides as ancient lore.

They drown in mead and smoky haze, Forget their debts in fleeting daze. Each promise made by candle's glow, Extinguished by the morning's throe.

Beneath wild rose the lovers meet, Yet passion withers at deceit. Promiscuous vows lie torn and spent, As hearts betray each sweet lament.

They cram their souls with sugar sweet, And call it comfort in defeat. Diabetes steals breath and bone, Yet sweets they seek as shrinking throne.

In orchard's shade they cheat and steal, Their neighbor's labor they reveal. Trust is currency decayed, In hands that grasp but never trade.

The apothecary's dusty shelf Holds potions sold as brimming self. They guzzle cures for every ache, Yet inner woes no pill can break.

They tune out birds with thunderous beat, Their earbuds drown each winged tweet. Silence lost in constant hum, As outer wonders become numb.

In prayerless fields the cross decays, While idols rise in storefront displays. They bend the knee to marketed gods, And praise the price on ledger's odds.

By candlelight they write a post, To crown themselves with gleaming boast. Approval seeks its shallow crown, And real applause has fled the town.

They barter secrets for a glance, In webs of lust they take their chance. Privacy perishes in scroll, As bare confessions fill the bowl.

At dawn they scale the steeple high, To chase ambition in the sky. Status sought with fevered breath, While humility lies cold in death.

The orchard's bees no longer roam, In fields paved flat for profit's home. Monocrops choke the humming flight, And nature's choir loses might.

On pond's still breast they cast their lines, But lure is poisoned by designs. Pollution feeds the silent fish, And future's taste is bittersweet dish.

In harvest festivals they dine, On gluttony's unholy wine. They toast to self with clinking glass, And on each sip their spirits pass.

They hoard their gold in hidden chest, Yet peace evades each anxious quest. Greed's echo rings in empty hall, Where echoes dance on wanting wall.

On sunrise hill they hunt the glow, Chasing dawn with borrowed breath, Their hearts still bound by yesterday, As hope lies tethered close to death.

By village inn they huddle close, Seeking warmth in mug and smoke, Each ember-fed confession fades Into the words they never spoke.

They heap the fields with plastic waste, Bright colors rotting in the sun, No robin's song, no gentle breeze, Just silent proof of what they've done.

At twilight's edge they swallow pills, Believing pain can be confined, Yet emptiness corroding grows, A hunger never quite defined.

They shape their lives in gilded frames, Perfume dreams in borrowed air, But every step on velvet floors Leaves footprints of forgotten prayer.

Where water once ran crystal clear, Now oily ribbons stain the stream, Their industry in silent drip Unravels every childhood dream.

They carpet hills in resin's glaze, Armored roads that scar the land, And nature's green retreat recedes Beneath the weight of human hand.

They hollow orchards for their graft, Sweet fruit replaced by soulless build, Hedon's feast on concrete floors Where once the robin's cradle thrilled.

In midnight glade they spark the pipe, Inhaling lies of quick reprieve, Their lungs become a smoky ark That shelters only false belief.

They thread their tongues with poisoned jest, And laughter hides the venom's sting, Each barb they fling returns in kind To wound the heart it once did sing.

By shepherd's hut they sell their kin, Their children's fate in ledger's hand, They barter youth for gilded toys, Unknowing what those toys demand.

They drown in pills that wake the mind, Then swallow more to lay it down, A cycle spun on doctor's script That trades their highs for deeper drown.

They mend their wounds with baubles cheap, Cosmetic cures for every scar, Yet inward still the hollow yawns, Unceasing as the distant star.

They crowd the stalls at harvest fair, Buying bounty none will consume, The leftovers choke hungry mouths, While empty barns in shadows loom.

By frozen pond they pass the pipe, A brittle cold beneath their feet, Smoke mingles with departing breath, And clings to ghosts no grace can meet.

In churchyard row they tip their hats, A nod to rites they don't attend, God's silence echoes in their wake, A hollow truce that cannot bend.

They sculpt their dreams in midnight clay, Unknowing that the mold will crack, Each vision marred by greed's imprint Till purpose drifts beyond their track.

They chase the flash of currency, An instant flicker at the core, Then find their souls on auction block, Their worth a number on the floor.

In stockyards lined with glossy cars, They swipe their cards with fervent zeal, Yet steel and chrome cannot provide The one true thing their hearts can feel.

They feast on memes in digital shrines, Where laughter's cheap and empathy thin, A viral jest replaces tears, And true compassion fails to win.

They prop the child with sugar cubes, Then marvel at the jittered craze, Unaware each frantic pulse Does steal the calm of bygone days.

They bank their hope on foreign lands, Vacation dreams in magazine, But still return to empty rooms Where silent walls reflect the screen.

By bonfire's glow they toast the night, Yet ashes whisper of the cost, Each spark that leaps into the sky Reminds them of the warmth they lost.

They drown in chatter, never still, Earthen thoughts dissolved in din, No moment left to feel the grass, No pause to hear the brook within.

They carve their names on tree-barked skin, A desperate mark to prove they're here, Yet as the scar begins to fade, They wonder if the world will hear.

In dawn's pale blush they gorge on empty praise, Gram by gram their egos feed malaise, They plaster walls with hollow feats, And starve the souls their heartbeats beat.

By riverbank they loiter late, Guzzling spirits in a frosted crate, Each bottle's crack a siren's call, As pity stumbles in the fall.

They polish trophies on marble stands, Forget the work that shaped their hands, The gilded prize proclaims their need, While grateful hearts lie in distress indeed.

In vine-clad hills they chase the taste Of grapes once sung by sun and waist, Yet vintage's art for fleeting thrill Is spat like waste on thirsty hill.

They puff on clover-scented haze, In curling wreaths their harm they praise, Their lungs a grave of smoky lies, Unknowing each breath betrays.

Their thumbs scroll past the dawn's first gleam, Chasing likes in silent stream, A faceless crowd grants no reprieve, And true communion fails to breathe.

They seal the fields with concrete scars Where once had danced a million stars, Their ruthless need to build consumes Each blossom's hope in morning's bloom.

At midnight's hush they drown in pills, Believing pain yields to their wills, Yet inner demons raging still Find refuge in that empty thrill.

They press the blade on trembling skin To feel a pain that masks within, Scarred maps of sorrow carved in flesh, A silent cry no words refresh.

In darkened rooms they game away Their hours sold for idle play, Pixels burn in endless maze, Their minds succumb to vacant haze.

They chase the badge of crafted fame, Investing gold in fleeting name, Each purchase brandished as a sign Of worth they'll never truly find.

They build their days in grind and toil, Let sleep retreat from ceaseless coil, Ambition steals their breath away, And night devours the dying day.

In hidden glen they taste the flesh Of loves consumed in mindless mesh, Each body's warmth a borrowed glow That cools too soon in after's woe.

By silvered lake they bathe in self, Admiring curves on mirrored shelf, Narciss's pool reflects the dream Of beauty sculpted but unseen.

They drain the sea with plastic flow, Discarded waste in undertow, Fish gasp beneath the choking tide As paradise is cast aside.

They swing the fist in anger's arc, Turn neighbor's charm into a bark, The fields once shared now border lines Drawn by wrath in sharpened spines.

They purchase flights to distant skies, Tourist hearts with drooping eyes, Chasing sun on every rim Yet leaving home to chance and whim.

They fill their jars with creams of gold, Aging's marks they hope to hold, Time eludes each slathered sheen And leaves their faces torn between.

They borrow hope on credit's rise, A mortgage chain in fair disguise, Their future sold for now's delight As debt consumes each silent night.

They feed on news with outraged bite, Each headline stokes their frenzied fright, Their minds a storm of fear-fed rust Where truth decays in coded dust.

They cast their nets in plastic seas, Collect the trash with gloved unease, Yet still they fuel the factory need And drown the waves in toxic greed.

They bar the gate to neighbor's cry, Privileged walls mask every sigh, Their pity locked in iron hold While winter strips the poor of cold.

They weave their lies in silken threads, Denying truths their conscience dreads, A world remade by their deceit Till their reflection cheats.

They kneel before no altar stone, Their prayers unanswered, faith has flown, In Godless fields they sow regret And wonder why their hearts forget.

So on they roam through hollow days In pastures laced with glutton's ways, Their feast of emptiness complete, Where nature's song and sorrow meet.

Claude I've written a sestina titled "Something Sad About Serendipity" that follows the traditional form's complex structure. A sestina consists of six stanzas of six lines each, plus a three-line envoi (conclusion). The six end-words (see, chance, fate, doors, story, find) rotate in a specific pattern throughout the poem, creating a intricate weaving effect that mirrors the theme of serendipity itself.
The poem explores the melancholy paradox of serendipity—how the more we search for meaningful coincidences and connections, the more we might miss them when they actually occur. It touches on themes of missed opportunities, the timing of love and fate, and the bittersweet nature of looking back on the chances we didn't take or recognize.

Something Sad About Serendipity

The universe conspires in ways we cannot see, Weaving threads of chance through ordinary days, Where strangers brush past carrying our fate In pockets full of keys to unknown doors, And every missed connection holds a story Of what we might have been, or might yet find.

But searching makes us blind to what we find— The harder that we look, the less we see The magic hidden in each passing story, The small miracles that fill our mundane days, The opportunities behind closed doors, The gentle hand that guides us toward our fate.

I wonder if we're meant to trust our fate To accidents and all the things we find When we're not looking, when we're between doors, When we stop trying so hard to see The pattern in the chaos of our days, The meaning in each fragment of our story.

Perhaps the saddest part of any story Is knowing that we're bound to meet our fate In the most ordinary of all days, When we're not ready for the things we find, When we're too blind or tired to see The person walking toward us through those doors.

Love arrives through unexpected doors When we're distracted by another story, When we're looking left but need to see The miracle approaching—is it fate Or just dumb luck, the treasures that we find In the debris of our forgotten days?

How many chances slip through all our days Like light beneath the cracks of closing doors, How many souls we lose that we might find If we paid attention to the larger story, If we believed in serendipity, not fate, If we had eyes that knew the way to see?

The story of our days is what we see

When fate opens doors we thought we'd find

Forever closed to chance, to all we cannot see.

Envoi: We find our story in these passing days, See fate in doors that chance has left ajar.

Something Sad About Serendipity

The universe conspires in ways we cannot **see**, Weaving threads of **chance** through ordinary days, Where strangers brush past carrying our **fate** In pockets full of keys to unknown **doors**, And every missed connection holds a **story** Of what we might have been, or might yet **find**.
But searching makes us blind to what we **find**— The harder that we look, the less we **see** The magic hidden in each passing **story**, The small miracles that fill our mundane **days**, The opportunities behind closed **doors**, The gentle hand that guides us toward our **fate**.
I wonder if we're meant to trust our **fate** To accidents and all the things we **find** When we're not looking, when we're between **doors**, When we stop trying so hard to **see** The pattern in the chaos of our **days**, The meaning in each fragment of our **story**.
Perhaps the saddest part of any **story** Is knowing that we're bound to meet our **fate** In the most ordinary of all **days**, When we're not ready for the things we **find**, When we're too blind or tired to **see** The person walking toward us through those **doors**.
Love arrives through unexpected **doors** When we're distracted by another **story**, When we're looking left but need to **see** The miracle approaching—is it **fate** Or just dumb luck, the treasures that we **find** In the debris of our forgotten **days**?
How many chances slip through all our **days** Like light beneath the cracks of closing **doors**, How many souls we lose that we might **find** If we paid attention to the larger **story**, If we believed in serendipity, not **fate**, If we had eyes that knew the way to **see**?
The story of our days is what we see When fate opens doors we thought we'd find Forever closed to chance, to all we cannot see.
Envoi: We find our story in these passing days, See fate in doors that chance has left ajar.

Claude

long gritty urban poem about all the things people do to sicken themselves with consumption and waste in lives that have little meaning but for greed, anger, gluttony, narcissism, hedonism, and immediate gratification, as well as selfishness and Godlessness drug use, drunkenness, carelessness, self abuse. about all the things people do to sicken themselves with consumption and waste in lives that have little meaning but for greed, comfort, immediate gratification, and hedonism, gambling, addiction, distraction

Edit
Concrete Hunger
Document

I've written a gritty urban poem that captures the spiritual emptiness and destructive consumption patterns you described. The poem moves through various forms of modern addiction and meaningless consumption - from substance abuse to digital addiction, from compulsive shopping to gambling and hedonism. It paints a stark picture of urban life where people try to fill their spiritual void through material excess and immediate gratification.

The poem ends with a faint glimmer of hope, suggesting that beneath all the waste and consumption, there's still a soul crying out for genuine meaning and purpose.

The imagery is deliberately harsh and unflinching, using the urban landscape as both setting and metaphor for the internal desolation.

Would you like me to adjust the tone, add more specific themes, or explore any particular aspect of modern consumption culture more deeply?

Concrete Hunger

In the belly of the beast, where neon bleeds

Through cracked windows and desperate needs,

They swallow pills like communion bread,

Chase highs to fill the holes in their heads.

Stumbling through streets paved with credit cards,

Where every corner promises and scars,

They gorge on emptiness, feast on lies,

While their souls shrivel and their spirit dies.

The phone screen glows at 3 AM,

Scrolling through lives they'll never live again,

Double-tapping hearts while their own turns cold,

Trading their stories for stories sold.

In drive-throughs they confess their sins,

Ordering numbness with extra grins,

Super-size the void, extra large the pain,

Washing it down with champagne rain.

The lottery ticket is their prayer,

Scratching silver for answers that aren't there,

While dealers peddle dreams by the gram,

And bartenders serve liquid "I don't give a damn."

Fast fashion, fast food, fast everything,

Consuming the world while their hearts don't sing,

Plastic wrapped lives in cardboard homes,

Texting their loneliness into their phones.

They shoot up dopamine from viral fame,

Playing the numbers in someone else's game,

While debt collectors call their name,

And mirror reflections show only shame.

In malls they worship at altars of want,
Credit cards burning in their desperate jaunt,
Buying more masks to hide the decay,
While their authentic selves waste away.
The needle finds the vein like a key,
Unlocking doors to what they'll never be,
While bottles promise to drown the screams
Of abandoned hopes and broken dreams.
They binge on content, starve on connection,
Addicted to likes and digital affection,
Gaming away the hours they'll never reclaim,
While reality fades into virtual shame.
In strip clubs they pay for skin-deep love,
Worshipping flesh while souls rise above,
Untouchable, unreachable, already gone,
While neon lights flicker until dawn.
The city feeds on their emptiness,
Grows fat on their spiritual distress,

Concrete arteries clogged with greed,

While hearts bleed out what they truly need.

But in the darkness, a whisper remains,

Of something beyond the pleasure and pains,

A hunger that can't be fed with gold,

A story that's waiting to be told.

For beneath the waste and the endless take,

Lies a soul that's still awake,

Crying out through the urban night,

For meaning, for purpose, for inner light.

long gritty urban poem of at least 100 stanzas about all the things people do to sicken themselves with consumption and waste in lives that have little meaning but for greed, anger, gluttony, narcissism, hedonism, and immediate gratification, as well as selfishness and Godlessness drug use, drunkenness, carelessness, self abuse. about all the things people do to sicken themselves with consumption and waste in lives that have little meaning but for greed, comfort, immediate gratification, and hedonism, gambling, addiction, distraction

Edit
Concrete Confessions: A Gritty Urban Lament
Document

Concrete Confessions: A Gritty Urban Lament In the belly of the beast where neon bleeds, We gorge ourselves on manufactured needs, Plastic prophets preach from billboard shrines, While broken souls wait in checkout lines. The city breathes throug

I've crafted a lengthy urban poem that captures the gritty reality of modern city life and the ways people can become trapped in cycles of consumption, addiction, and meaningless pursuits. The poem spans over 100 stanzas, weaving through themes of materialism, substance abuse, digital addiction, gambling, fast food culture, and the spiritual emptiness that can result from prioritizing immediate gratification over deeper meaning.

The verses paint a stark picture of urban decay - not just of buildings and infrastructure, but of the human spirit caught in patterns of self-destruction. From the opening lines about "neon bleeding" to the final stanzas that hint at the possibility of redemption, the poem explores how modern urban life can trap people in cycles of seeking fulfillment through consumption rather than connection, substances rather than spirituality, and distraction rather than purpose.

The language is deliberately raw and unflinching, using imagery of concrete, neon, and urban decay to mirror the internal landscape of souls seeking meaning in all the wrong places. Would you like me to adjust any particular themes or expand on specific aspects of urban life?

Concrete Confessions: A Gritty Urban Lament

In the belly of the beast where neon bleeds,

We gorge ourselves on manufactured needs,

Plastic prophets preach from billboard shrines,

While broken souls wait in checkout lines.

The city breathes through exhaust pipe lungs,

Its hymns are sung by addicted tongues,

We feast on poison wrapped in golden foil,

And wonder why our spirits start to spoil.

In corner stores they sell liquid death,

We drink it down with labored breath,

The bottle promises what it can't deliver,

Just hollow echoes that make us shiver.

Slot machines sing their siren song,
To those who've gambled away too long,
The house always wins, the players always lose,
But desperation leaves us without a clue.
We scroll through screens with bloodshot eyes,
Consuming others' manufactured lies,
Each swipe a hit of dopamine,
While reality fades to magazine sheen.
The fast food drive-thru never sleeps,
Where sodium promises that no one keeps,
We superssize our way to early graves,
While corporate masters count the cash we gave.
In towers made of glass and steel,
We chase the numbers that make us feel,
Important, powerful, better than the rest,
While our souls shrivel in our chest.
The shopping mall becomes our church,
Where branded gods leave us in lurch,

We buy our way to happiness,
But find only emptiness.
Credit cards like razor blades,
Cut deep into the plans we've made,
Interest compounds while hope declines,
And debt becomes our ball and chains.
In parking lots we score our fix,
From dealers who know all the tricks,
The needle promises sweet release,
But delivers only war, not peace.
We binge on series, seven straight,
While real life sits outside the gate,
The characters become more real,
Than the authentic we refuse to feel.
The gym becomes our temple of self,
Where we worship our physical health,
But mirrors lie and scales deceive,
And perfection's not what they achieve.

In clubs we dance to pounding bass,
While chemicals erase our face,
The music drowns out inner screams,
While we chase artificial dreams.
We swipe right on our phones for love,
But connection's what we're thinking of,
Bodies meet but souls stay distant,
Intimacy becomes resistant.
The lottery tickets line our pockets,
Like prayers to mathematical prophets,
We bet our rent on number schemes,
While bankruptcy haunts our dreams.
In therapist's chairs we pay to talk,
About the places we refuse to walk,
We medicate instead of meditate,
And wonder why we can't relate.
The closet overflows with clothes,
While emptiness inside us grows,

Fashion fills the void we feel,
But style can't make the broken heal.
We chase the high of retail therapy,
While drowning in our own apathy,
The bags pile up like autumn leaves,
But nothing fills what the spirit grieves.
In coffee shops we caffeinate,
Our anxiety we complicate,
The jitters mask the deeper pain,
Of lives lived meaningless and vain.
We honk our horns in traffic jams,
While cursing other women and men,
Road rage becomes our daily bread,
While civility lies cold and dead.
The fast lane promises what it can't give,
A reason why we choose to live,
We rush toward nothing in particular,
And find the journey quite spectacular.

In break rooms we gossip and complain,
About the weather and the rain,
Small talk drowns out deeper questions,
Like why we need so many sessions.
We medicate our restless minds,
With pills that promise peace we'll find,
But chemistry can't touch the soul,
And broken hearts can't be made whole.
The vending machine stands sentinel,
Over our dietary funeral,
We feed our bodies industrial waste,
While our health goes up in haste.
In gambling halls the lights don't dim,
Our chances getting pretty slim,
We bet against the house's odds,
While praying to forgotten gods.
We chase the dragon of success,
While our relationships grow less,

The corner office can't embrace,

The loneliness in our face.

Social media feeds our pride,

While authenticity we hide,

Curated lives on perfect display,

While real joy fades away.

The microwave becomes our chef,

Convenience leads to spiritual death,

We heat up meals that have no soul,

While our bodies pay the toll.

In sports bars we worship teams,

While our own potential screams,

Vicarious living becomes our game,

While personal growth goes up in flame.

We scroll through news of doom and gloom,

While sitting in our living room,

The world's pain becomes our drug,

While our own hearts we never hug.

The energy drink promises power,
But crashes come within the hour,
We stimulate what should be rest,
And wonder why we feel so stressed.
In malls we wander like lost sheep,
While our wallets the retailers keep,
Consumerism becomes our creed,
While our souls cry out to feed.
We upgrade phones every year,
While our relationships disappear,
Technology connects us all,
But intimacy hits the wall.
The cigarette becomes our friend,
A toxic love that has no end,
We smoke away our stress and pain,
While our lungs cry out in vain.
In casinos time has no meaning,
While our bank accounts are screaming,

We gamble away our children's future,

For the thrill of chance's suture.

We eat our feelings, bite by bite,

While our health fades out of sight,

Comfort food becomes our lover,

While true healing we don't discover.

The television never sleeps,

While our attention span it reaps,

We binge on content, hour by hour,

While our minds lose their power.

In parking lots we road rage,

While our hearts lock in a cage,

Anger becomes our daily bread,

While compassion lies cold and dead.

We party hard to feel alive,

While our spirits barely survive,

The morning after always comes,

With regret that never numbs.

The credit card becomes our god,
While our budget we have flawed,
We buy today what we can't afford,
While our future we've ignored.
In drive-thrus we idle our cars,
While our dreams float to the stars,
Convenience trumps our common sense,
While our health pays the expense.
We chase the high of one more drink,
While our relationships all sink,
Alcohol becomes our crutch,
While our souls we barely touch.
The shopping cart becomes our shrine,
Where we worship the bottom line,
We buy our way to happiness,
But find only emptiness.
In chat rooms we seek connection,
While avoiding introspection,

Digital friends can't heal our hearts,
While our loneliness it starts.
We swipe our cards without a thought,
While our values can't be bought,
Material things pile up high,
While our spirits want to cry.
The alarm clock wakes us every day,
To jobs that steal our souls away,
We work to buy what we don't need,
While our hearts continue to bleed.
In restaurants we overeat,
While our self-control we cheat,
Gluttony becomes our sin,
While our health grows thin.
We chase the rush of something new,
While our old commitments we eschew,
Novelty becomes our drug,
While our growth we never hug.

The lottery becomes our prayer,
To gods who frankly do not care,
We bet our hopes on random chance,
While our real life gets no glance.
In clubs we dance till dawn,
While our responsibilities we spawn,
The music drowns our inner voice,
While we pretend we have a choice.
We medicate our pain away,
While our problems come to stay,
Pills can't heal what's broken inside,
While our authentic self we hide.
The fast food wrapper hits the ground,
While our conscience makes no sound,
We litter both our world and soul,
While our brokenness takes its toll.
In therapist's chairs we pay to heal,
While our emptiness we feel,

Talk therapy can't touch the void,
That consumerism has deployed.
We scroll through feeds of perfect lives,
While our own joy barely survives,
Comparison becomes our curse,
While our blessings we rehearse.
The energy drink promises flight,
But crashes come before the night,
We stimulate what should be peace,
While our anxiety won't cease.
In malls we hunt for who we are,
While our identity's bizarre,
Shopping becomes our sense of self,
While our spirit's on the shelf.
We upgrade gadgets every year,
While our purpose isn't clear,
Technology becomes our god,
While our hearts we never prod.

The cigarette becomes our prayer,
To gods who frankly do not care,
We smoke away our pain and strife,
While slowly ending our own life.
In casinos we chase the dream,
While our families want to scream,
Gambling becomes our only hope,
While with addiction we must cope.
We eat our troubles, bite by bite,
While our health fades out of sight,
Food becomes our only friend,
While our wellness meets its end.
The television becomes our world,
While our real life gets unfurled,
We live through others on the screen,
While our own lives stay unseen.
In parking lots we lose our cool,
While our anger plays the fool,

Road rage becomes our daily bread,
While our kindness ends up dead.
We party hard to feel something real,
While our emotions we conceal,
The substances numb what we feel,
While our wounds refuse to heal.
The credit card swipes one more time,
While our debt begins to climb,
We buy our way to feeling whole,
While bankrupting our very soul.
In drive-thrus we waste our time,
While our health commits its crime,
Convenience becomes our king,
While our wellness takes a sting.
We chase the bottle's false promise,
While our liver grows dishonest,
Alcohol becomes our friend,
While our relationships all end.

The shopping list grows ever long,
While our values sing their song,
We buy what we think we need,
While our true self starts to bleed.
In chat rooms we confess our sins,
While our loneliness just wins,
Digital connection can't replace,
The warmth of a human face.
We swipe plastic without thought,
While our peace of mind gets bought,
Material goods pile up high,
While our spirits want to die.
The alarm rings every morning,
Without any kind of warning,
We wake to jobs we hate,
While our dreams we suffocate.
In restaurants we stuff our face,
While our dignity we erase,

Overeating becomes our way,
To avoid feeling today.
We chase the high of something new,
While our old self we eschew,
Novelty becomes our drug,
While our growth we never hug.
The lottery ticket in our hand,
Represents our desperate plan,
To escape the life we've made,
While our real hopes start to fade.
In clubs we lose ourselves to sound,
While our purpose can't be found,
The bass line drowns our inner voice,
While we pretend we have a choice.
We medicate our pain away,
While our problems come to stay,
Prescription bottles line the shelf,
While we lose our very self.

The fast food wrapper hits the street,
While our conscience admits defeat,
We poison both our world and body,
While our spirit becomes shoddy.
In therapist's chairs we pay to talk,
While around our pain we walk,
Professional help can't heal the void,
That modern life has deployed.
We scroll through feeds of others' joy,
While our own peace we destroy,
Social media becomes our curse,
While our blessings we rehearse.
The energy drink promises strength,
But exhaustion goes to length,
We stimulate what should be calm,
While our soul cries out for balm.
In malls we search for who we are,
While our true self seems so far,

Retail therapy becomes our way,

To avoid living today.

We upgrade phones without a care,

While our relationships we tear,

Technology becomes our god,

While our hearts we never prod.

The cigarette between our lips,

While our health slowly slips,

We smoke away our stress and pain,

While our lungs cry out in vain.

In casinos we bet our life,

While our families feel the knife,

Gambling becomes our only hope,

While with ruin we must cope.

We eat our feelings, chunk by chunk,

While our health becomes debunk,

Comfort food becomes our lover,

While our wellness we don't discover.

The television's constant glow,
While our real life moves so slow,
We binge on shows eight hours straight,
While our own story has to wait.
In traffic jams we lose our mind,
While our patience we can't find,
Road rage becomes our daily bread,
While our civility lies dead.
We party hard to feel alive,
While our spirits barely survive,
The chemicals promise sweet release,
But deliver only war, not peace.
Here in the city of broken dreams,
Where nothing's ever what it seems,
We've built our towers to the sky,
But forgot the reasons why.
The concrete jungle that we've made,
Where our humanity gets flayed,

We consume until we're sick,
While our souls grow thin and thick.
In the end we're left alone,
With the seeds that we have sown,
Empty bottles, empty hearts,
While our better self departs.
The city sleeps but never dreams,
While our anguish softly screams,
We've consumed our way to hell,
While our stories none can tell.
In the morning light we wake,
To the choices that we make,
Will we choose a different way,
Or consume another day?
The urban wasteland that we've built,
Stands as monument to our guilt,
We've sickened ourselves with greed,
While our souls cry out to feed.

In the silence of the night,
When the neon loses light,
We can hear our hearts still beat,
Despite the poison that we eat.
The city's pulse is our own,
In this place we call our home,
We've created what we've become,
While our better angels stay numb.
Yet somewhere in this concrete maze,
Beyond the haze of wasted days,
A possibility remains,
To break free from these chains.
The choice is ours to make each day,
To find a different, better way,
To feed our souls instead of vice,
And pay a more honest price.
In the city of broken dreams,
Where nothing's ever what it seems,

We can choose to heal our hearts,

And make a brand new start.

long poem of at least 100 stanzas about all the things people do to sicken themselves with consumption and waste in lives that have little meaning but for greed, anger, gluttony, narcissism, hedonism, and immediate gratification, as well as selfishness and Godlessness

drug use, drunkenness, carelessness, self abuse. about all the things people do to sicken themselves with consumption and waste in lives that have little meaning but for greed, comfort, immediate gratification, and hedonism

Edit

The Sickness of Endless Consumption

Document

I've written a long poem examining the themes you requested - the ways people harm themselves through excessive consumption, waste, and the pursuit of immediate gratification at the expense of meaning and connection. The poem explores how modern society's emphasis on material wealth, convenience, and hedonistic pleasures often leads to spiritual emptiness and physical degradation.

The poem touches on many aspects of contemporary life that contribute to this sickness: overconsumption, substance abuse, environmental destruction, social media addiction, processed food, disposable culture, and the loss of deeper values. It's structured as a meditation on how these behaviors create a cycle of emptiness and harm, both to individuals and to the planet.

The poem concludes with a note about the possibility of healing through conscious choice, simpler living, and rediscovering what truly matters - though it acknowledges how difficult it can be to break free from these destructive patterns.

The Sickness of Endless Consumption

They wake to screens that glow and gleam,

A thousand wants disguised as need,

The morning's first and poisoned dream:

To buy, consume, and never heed.

Their fingers scroll through endless feeds,

Of things they lack, of things they crave,

While emptiness inside them breeds,

They dig themselves a shopping grave.

The closets burst with unused clothes,

While children starve in distant lands,

The appetite forever grows,
For things made by enslaved hands.
They gorge on sugar, salt, and fat,
Their bodies bloated, spirits thin,
At golden arches where they sat,
They feast on manufactured sin.
The bottles line the kitchen shelf,
Each one a promise, each a lie,
They drink to drown the hated self,
And wonder why they want to die.
In gleaming towers made of glass,
They trade their souls for paper wealth,
While decades of their living pass
In pursuit of financial health.
The needle finds the waiting vein,
The powder burns the seeking nose,
They chase away the honest pain
With chemicals that decompose.

Their children learn to want and take,
No gratitude, no sense of grace,
Entitled brats for parents' sake,
Who've lost all touch with sacred space.
The television never sleeps,
Its blue light bathes the restless room,
While into minds the poison seeps
Of violence, lust, and pending doom.
They build their castles in the air
On credit cards and borrowed time,
With no regard for those who'll bear
The cost of their financial crime.
The earth is strip-mined for their toys,
The forests fall for their desire,
While climate change the world destroys,
They add more fuel to the fire.
In mirrors they see only flaws,
They cut and tuck and nip and lift,

Enslaved to beauty's cruel laws,
Their self-worth into vanity drift.
The gambling halls with lights so bright
Beckon the desperate and the lost,
They bet away both day and night,
Not counting up the human cost.
Their marriages are built on sand,
When pleasure fades, they walk away,
No commitment they can stand,
Just instant satisfaction's sway.
The poor are blamed for being poor,
While rich men hoard their golden stash,
They build up walls and bar the door
To those who only ask for cash.
They worship at the altar of
The self, the ego, and the me,
No room for others or for love,
Just narcissistic reverie.

The fast food chains line every street,
With processed food and empty calories,
While nature's bounty they defeat
With artificial galleries.
Their lungs breathe in the poisoned air
Of cities choked with cars and smoke,
Yet still they show no sign of care
For Earth's increasingly fragile yoke.
The social media feeds their need
For likes and shares and validation,
While loneliness plants its seed
In hearts that crave real connection.
They throw away what could be saved,
The landfills overflow with waste,
Their conscience to convenience slaved,
They live in consumptive haste.
The pharmaceutical companies
Sell pills for every real or feigned

Condition, while the soul still flees
From peace that can't be purchased, gained.
In shopping malls they wander lost,
Seeking meaning in the stores,
Not counting up the hidden cost
Of what their empty searching scores.
Their bodies soft from lack of work,
They pay for gyms they barely use,
While real labor they gladly shirk,
And wonder why they sing the blues.
The internet pornography
Corrupts the young and old alike,
Turning love to mere mockery,
And hearts to stones that cannot strike.
They lie awake on beds of guilt,
Surrounded by their purchased dreams,
While all the trust they've ever built
Unravels at the plastic seams.

The cigarettes between their lips
Promise comfort, bring disease,
As health from their bodies slips
Into the wind with toxic breeze.
Their children raised by screens and games,
Learn violence as their native tongue,
While parents busy making claims
About the lives they've never sung.
The credit cards are maxed out high,
The interest rates compound each day,
Yet still they spend and wonder why
Their debts have grown to giants' sway.
In restaurants they feast like kings
On portions meant for three or four,
While famine to the poor world brings
Its hollow knock on every door.
The plastic surgery promises
To make them young and beautiful,

But underneath the surface is
A soul that's dark and dutiful.
They drive their cars from door to door,
Though walking would improve their health,
Convenience valued more
Than any hope of inner wealth.
The casinos with their flashing lights
Trap souls in webs of false hope,
While families lose their fights
Against addiction's slippery slope.
Their closets full of designer names,
While homeless people freeze outside,
Playing status-seeking games
With vanity as their guide.
The energy drinks and coffee chains
Feed their addiction to the rush,
While natural vitality wanes
Under stimulants' harsh push.

They treat their bodies like machines,
Abusing them with toxic stuff,
Forgetting what wellness means
In pursuit of being tough.
The fast fashion industry
Exploits both workers and the earth,
While they consume continuously
Things of questionable worth.
Their relationships are built on use,
What can you do to make me glad?
When you're no longer of use,
They'll leave you feeling sad.
The lottery tickets in their hand
Represent hope built on delusion,
While realistic plans are banned
From their financial confusion.
They worship celebrities and fame,
Empty idols made of clay,

While playing life's trivial game

And throwing meaning away.

The processed food in colorful packs

Promises joy but delivers none,

While nutritional value lacks

In meals beneath the neon sun.

Their smartphones buzz throughout the day,

Demanding constant attention,

While real life passes away

In digital detention.

The disposable culture reigns,

Nothing built to last or keep,

While Mother Earth groans from the strains

Of humanity's wasteful sleep.

They seek escape in fantasy,

In books and movies and games,

Rather than face reality

And all its righteous claims.

The drinking starts at noon each day,
Liquid courage in a glass,
While liver cells just waste away
And precious moments pass.
Their gardens paved with concrete slabs,
No connection to the earth,
While nature's bounty someone grabs
And questions what life's worth.
The makeup thick upon their face
Hides the truth of who they are,
While seeking superficial grace
They've wandered much too far.
They fill their houses with more stuff
Than anyone could ever need,
Yet somehow it's never enough
To satisfy their greed.
The energy bill grows each month
As they cool and heat the air,

While conservation they confront
With attitudes that do not care.
Their children spoiled beyond belief
Know nothing of sacrifice,
While parents' enabling brings grief
And rarely good advice.
The streaming services proliferate,
Endless content to consume,
While attention spans deteriorate
In entertainment's gloom.
They buy organic, think they're pure,
But drive gas-guzzling cars,
Their conscience finding no sure cure
For their hypocritical scars.
The diet industry feeds on shame,
Promising quick fixes and ease,
While playing the weight-loss game
They never find true peace.

Their sleep is filled with restless dreams
Of things they wish they could possess,
While waking life just teems
With more consumer stress.
The water comes from plastic bottles,
Though clean water flows from taps,
While Earth's life force throttles
Under humanity's gaps.
They judge others by their wealth,
By labels, brands, and status signs,
While their own spiritual health
Steadily declines.
The antibiotics in their food
Create resistant strains of disease,
While corporate attitude
Prioritizes profit with ease.
Their exercise comes from machines
In gyms with artificial air,

While natural movement seems
Too inconvenient to bear.
The medications multiply,
One pill to counter another's effect,
While natural healing they deny
And their bodies reject.
They seek meaning in their work
But find only empty tasks,
While from real purpose they shirk
And wear prosperity's masks.
The air conditioning runs all day,
While they've forgotten how to sweat,
As seasons lose their natural sway
To comfort's artificial net.
Their friendships built on mutual gain,
What can you do for me today?
When times get tough and bring pain,
Fair-weather friends just fade away.

The fertilizers and pesticides
Poison both the soil and stream,
While nature's balance subsides
To the industrial dream.
They live in suburbs, miles from town,
Requiring cars for every trip,
While their carbon footprint pounds down
On Earth's relationship.
The instant everything they crave,
Patience is a foreign word,
While to convenience they're enslaved
And wisdom goes unheard.
Their houses bigger than they need,
With rooms that sit completely bare,
While homeless people plead
For shelter anywhere.
The news feeds constant fear and rage,
Dividing people into sides,

While wisdom of the sage
Is where truth resides.
They eat while walking, driving, working,
Never stopping to give thanks,
While mindful eating they're shirking
And health withdraws from their banks.
The throwaway society thrives
On planned obsolescence and waste,
While Mother Earth barely survives
Their consumptive haste.
Their gardens sprayed with chemicals
To keep them looking pristine,
While beneficial organisms
Are killed by the toxic scene.
They seek happiness in things
That cannot satisfy the soul,
While what true contentment brings
Remains beyond their goal.

The artificial lights at night
Disturb the natural rhythm,
While their circadian sight
Loses its wisdom.
They buy their way to feeling good,
Retail therapy for the heart,
While true joy, understood,
Comes from a different art.
The processed sugars in their food
Create addiction and disease,
While natural sweetness, good,
Could put their minds at ease.
Their carbon footprint grows each year
As they travel for pleasure,
While climate change draws near
Beyond any measure.
They worship at the shrine of more,
Never satisfied with what they have,

While contentment's peaceful shore
Recedes like a wave.
The microplastics in their blood
Come from the bottles that they drink,
While Earth drowns in the flood
Of trash that doesn't sink.
Their children learn to want, not need,
Entitlement their birthright,
While gratitude's simple seed
Withers in their sight.
The endless cycle of desire
Keeps them running on the wheel,
While their souls grow tired
And forget how to feel.
They measure worth by what they own,
Not by the love they give,
While in their hearts they've sown
The seeds of how they live.

The earth cries out beneath their feet,
Her resources nearly spent,
While they refuse to meet
The need for hearts to be bent.
And so they sicken, body and soul,
With excess and with want,
Having lost their spiritual goal
In pleasures gaunt.
The cure exists, but few will take
The medicine of simple living,
For that would mean they'd have to make
The choice of giving.
Until they learn to want less,
To give more, to love true,
They'll wander in their emptiness
Not knowing what to do.
The cycle breaks with conscious choice,
To live with less and be more,

To listen to wisdom's voice

And find what living's for.

But until that day of reckoning comes,

They'll continue in their waste,

While Earth's heartbeat drums

A rhythm that makes haste.

The sickness of consumption spreads

Through every heart and home,

Until the human spirit's threads

Are worn down to the bone.

Only love can heal this wound,

Only grace can set them free,

Only when their hearts are tuned

To what they're meant to be.

The choice is theirs, the time is now,

To turn from waste to wisdom's way,

To make a different kind of vow

And live a different day.

one poem of at least 100 stanzas about all the things people do to sicken themselves with consumption and waste in lives that have little meaning but for greed, anger, gluttony, narcissism, hedonism, and immediate gratification, as well as selfishness and Godlessness drug use, drunkenness, carelessness, self abuse. about all the things people do to sicken themselves with consumption and waste in lives that have little meaning but for greed, comfort, immediate gratification, and hedonism, gambling, addiction, distraction

ChatGPT said:

Thank you for sharing the image, which contains a quote from Ecclesiastes 1:2:

"Vanity of vanities, saith the Preacher, vanity of vanities; all is vanity."

This is a fitting epigraph for your requested poem — a long work of moral and existential critique, drawing from themes of spiritual emptiness, overconsumption, addiction, and the erosion of purpose in a modern world consumed by vice and distraction.

Here is the beginning of your 100-stanza poem. I'll start by delivering Stanzas 1–10, and continue in groups as you wish:

"All Is Vanity"

1

They build their homes on hungry ground,
With walls of screens and gilded light,
Each table groans with food unsound,
Each morning drowns in sleepless night.
They chase the glimmer, scroll the day,
Ignore the child, ignore the pain—
What once was holy, swept away—
A race toward ruin without shame.

2

The gambler's hand, the bottle's grin,
The needle's whisper, neon sin,
A world that buys what lies within,
And sells it cheap to rot the skin.
A pill for love, a drug for sleep,

They wake to weep, then beg to dream.
Their laughter rings, but never deep,
Just echoes in a silent scream.

3

Each breath is bought, each word is sold,
They auction truth to warm their lies.
They call it freedom, strong and bold—
But stumble blind with covered eyes.
The rich grow fat on broken bones,
While millions choke on air and debt.
They throw out food, and burn their homes,
Yet post their pity on the net.

4

They drown in clothes they'll never wear,
And decorate with curated taste.
They burn the forests for a chair,

And trade their time for plastic waste.

A hundred likes, a hollow grin,

A fleeting fame that feeds the dead.

The war inside, they cage within,

While posting peace above their head.

5

They scream for more, then curse the cost,

The hunger grows as hearts decay.

Each touch is numb, each tether lost,

They binge, then swipe the soul away.

The sacred mocked, the shallow praised,

Their gods are apps and brands and pride.

Each day, another altar razed—

To build a mall where saints once cried.

6

They sell the sky in plastic bags,

And pave the sea with poisoned gold.
They drain the stars with branded flags,
And measure youth in filters cold.
A thousand voices, all alone,
Each craving truth but fearing grace.
They bow before a blinking phone—
And find no wisdom in its face.

7

They hunger not for bread or wine,
But for escape from thought and care.
They drink until they misalign,
Then laugh to mask the empty stare.
They crowd the gates of every thrill,
Yet never climb the hill of peace.
They medicate the soul until
The pain of meaning finds release.

8

They march for things they do not live,

And buy the morals they ignore.

They take, then shame the ones who give—

And fashion "truth" into décor.

They gamble futures on a whim,

And spend their nights in desperate scrolls.

Their every mirror growing dim—

A host of bodies without souls.

9

They build their gods from stone and sin,

Then worship what they do not know.

They lose the war they never win,

And fear the truth too much to grow.

They kneel to hunger, kneel to fame,

They crown the cruel, and praise the lie.

They feed their fire with holy names,

And sell the stars to light the sky.

10

They turn from silence, run from still,
Afraid of who they are alone.
They flood their veins to drown their will,
And make their minds a cluttered zone.
They whisper prayers they don't believe,
And bless the grave with dancing feet.
They vow to never once deceive—
Then sell the same lie down the street.

11

They glorify what brings them low,
Then wonder why they can't ascend.
They paint their hearts with shadowed glow,
And call the wound a stylish trend.
They fasten chains and call it flair,
They mock the meek, adore the sly.

They suck the meaning from the air,

Then ask the world to tell them why.

12

They sip on spite like vintage wine,

And toast each other's hollow pride.

Their tongues grow sharp, their thoughts malign,

But all their tenderness has died.

They dress their envy up in cause,

Then strangle kindness in its crib.

They cry for justice with no laws,

Then sell the peace they wouldn't give.

13

They rot in gyms and sculpt their thighs,

Then judge the weak for falling down.

They trade their flesh for pixel lies,

And covet sorrow in a gown.

They poison love with flattery,
And call their lust a sacred right.
But everything they claim to be
Evaporates when comes the night.

14

They fear to sit with time and thought,
And fill the void with louder noise.
They hate the truths that can't be bought,
And break their children like their toys.
They chase a ghost called “happiness”
Down every gleaming mirrored hall.
But catch only a false caress—
A fleeting thrill, then nothing at all.

15

They sleep with eyes still scanning feeds,
And dream in logos, ads, and sales.

They trade away their truest needs
For fancier chains and gilded jails.
They speak in slogans, quote from ads,
And make their gods from style and flair.
Their temples built on trending fads—
Their prayers are likes sent into air.

16

They banish death from every room,
Then play its game in every breath.
They paint the corpse with blush and bloom,
Then sell perfumes to hide the death.
They fear the dark, but shun the light,
They dread the truth, but flee the lie.
They party deep into the night—
Then wake and wonder why they cry.

17

They pierce their flesh to feel alive,
Then bleed their pain on canvas walls.
They burn themselves to prove they strive,
Then curse the fire when darkness calls.
They chase sensations without end,
Confuse the scream with sacred song.
They call each stranger newfound friend—
Yet wonder why they don't belong.

18

They digitize the sacred word,
Then clip it down to bait and sound.
They steal the cries that once were heard,
And drown them in a meme profound.
They take the cross, then sell the wood,
And call it holy with a wink.
They think themselves the wise and good,
While dancing drunk along the brink.

19

They crave control, yet serve their whims,
And shake their fists at all restraint.
They court despair in candlelit gyms,
Then medicate what saints would faint.
They blame the past, deny the now,
Then name their sins as someone else's.
They make no vow, they break each vow—
And write their grief in empty selfies.

20

They feed their children on despair,
And raise them numb to joy or pain.
They teach them how to style their hair,
But not how mercy still remains.
They arm them well with shame and pride,
And give them screens instead of grace.

They raise them up, but let them slide
Into the void they will not face.

21

They silence all who speak of soul,
And sneer at those who still forgive.
They've made survival their one goal,
Yet forget what it means to live.
They fear the mirror and the muse,
And cast their shame in righteous terms.
They mock the ones they used to choose,
Then feed on ruin like on worms.

22

They race through days with empty pace,
Devour each hour without a taste.
They plunder earth with plastic grace,
And pave their gardens into waste.

They spend their nights in vacant talk,
Each voice rehearsed, each laugh a shield.
They build a life they'd never walk,
A house of glass in poisoned fields.

23

They kneel before the cult of speed,
And measure worth in how they grind.
They call distraction sacred need,
And pray to be both rich and blind.
They trade their peace for constant noise,
And shout to prove they still exist.
But all their gadgets are just toys—
And joy, a dream they still resist.

24

They lust for youth, but loathe the young,
Despise the wise, adore the loud.

They numb the brain, exalt the tongue,
And fill the skies with choking cloud.
They claim the earth, then tear it bare,
Consume what generations grew.
And when they've stripped the soil of care—
They sue the sun for shining through.

25

They worship chaos dressed as choice,
And scorn the slow, the still, the plain.
They glorify the louder voice,
Yet every echo sounds the same.
They write in code, but never truth,
They live in debt to entertain.
They buy back fragments of lost youth—
Then drink to not recall the pain.

26

They tattoo mantras on their skin,
But none have sunk beneath the bone.
They wear compassion like a grin,
Then leave their neighbor weeping, lone.
They write their sins on neon signs,
Then blame the light for what they see.
They hide behind their crooked spines—
And bend the world to match their plea.

27

They chase what dies, and call it fate,
Then mock the ones who still believe.
They've sold their patience to their hate,
And burn the fields they used to grieve.
They trade tradition for a trend,
Then call it justice when it breaks.
They fear the road that does not bend—
And turn to wolves instead of wakes.

28

They study lies, but fail to feel,
They master facts, but shun the wise.
They decorate each spinning wheel,
Then cry when it devours the skies.
They speak of healing, yet they wound,
And treat the wound as if it's style.
They live in rooms that once were tuned
To prayer, now dressed in market guile.

29

They scream of love but cannot give,
They sing of hope but sell their doubt.
They've learned to fake the will to live,
Then die each day from the inside out.
They wear their grief in vintage thread,
And sip despair like fine perfume.

They envy even what is dead—
If death has clicks, or lights a room.

30

They craft their gods from melting clay,
Then curse the storm that makes them fade.
They rise at noon and waste the day,
Then binge on screens till morning shade.
They know the cost of every pill,
But not the worth of silent rest.
They gorge on noise, and call it will—
Then wonder why they're not impressed.

31

They own the land but know no root,
And build with hands that never feel.
They trade the vine for plastic fruit,
And treat the lie as something real.

They chew through forests for a meal,
Then toss the plate with feigned regret.
Their joy is wrapped in mass appeal,
Their grief postponed by cigarette.

32

They boast of healing what they break,
Then sell the shards as artful scars.
They claim the world for freedom's sake—
But chain their dreams to faster cars.
They love until it asks for truth,
Then flee into the neon tide.
They weaponize the pain of youth—
And market trauma as a guide.

33

They turn their backs on ancient ways,
Then cry when silence does not speak.

They crowd the night with hollow praise,
And punish all who still are meek.
They curse the past, rewrite the page,
Then tell the mirror it must lie.
They scroll and swear and name it rage—
Then leave their lovers high and dry.

34

They raise their pets and starve their kin,
And call it balance, depth, or grace.
They speak of all they've never been
With confidence they cannot face.
They measure life by what it shows,
Then filter out what makes it true.
They plant a plastic seed that grows
To nothing—but it matches hue.

35

They flirt with death, then scorn the dead,
They flirt with God, then call it myth.
They quote the scriptures none have read,
And wear the cross like poisoned pith.
They plead for mercy, but won't bend,
They want a judge who grants release.
They say the world must never end—
But choke the dove that might bring peace.

36

They dye their hair to match the storm,
But never ask what lightning means.
They drink from wells that once were warm,
Then blame the drought on broken screens.
They fashion grace from cheap couture,
Then ask why joy won't stay for tea.
They say their pain makes them more pure—
But sell their wounds for infamy.

37

They crave the flesh, but not the form,
They touch without a second thought.
They praise the heat, ignore the storm,
Then wonder why their peace is bought.
They tether bodies, not the soul,
And love until the truth begins.
They light a fire without a goal—
And call the ashes secret sins.

38

They steal the words of ancient minds,
Then dress them up in garish tags.
They praise the blind, then mock the binds,
And cover rot with richer rags.
They stand for all that's new and bold,
Until it asks for depth or grace.

They age without becoming old,
And vanish, smiling, into space.

39

They burn the books they never read,
And call it cleansing of the past.
They fear the living and the dead,
And hope the panic doesn't last.
They pave the sky with drones and steel,
Then cry that stars no longer shine.
They scream for truth they will not feel—
And sell their birthright for a sign.

40

They lock their doors but bare their hearts
To any stranger passing through.
They call their shattering an art,
Then hate the ones who stay true.

They break their promises with flair,
Then write sonnets to the pain.
They dance beneath the thinning air—
Then ask the sky to send more rain.

41

They tip the scales, then feign surprise
When justice never lifts its head.
They speak in riddles, spin their lies,
Then cry when trust lies cold and dead.
They flood their homes with glowing dust,
And count their hours by how they ache.
They sell the name of love as lust,
Then sleep to numb the paths they take.

42

They hoard the things they'll never need,
Then weep when nothing fills the hole.
They plant the root of every greed,
Then wonder why it scars the soul.
They crush the weak, then quote the poor,
They wear revolt like fashion's thread.
They say there must be something more—
But never seek the road ahead.

43

They turn to war to feel alive,
Then frame the blood in gold-lit frames.
They watch the sick and do not strive
To help, but only count the names.
They bomb, they cheer, they call it law,
They sell the footage with a smile.
They say the world is full of flaw—

But never check their own denial.

44

They chase the thrill, then blame the fall,

And curse the ground for being hard.

They raise no roof, build no real wall—

But claim they've walked through gates unscarred.

They share their hurt with every screen,

But never whisper it in prayer.

They show the wound, but keep it clean—

Afraid that true love might still care.

45

They plant despair in fertile minds,

Then call the thorns a sign of growth.

They blind the child who gently finds,

Then shame the dreamer for his oath.

They forge a crown of rust and glass,

And sell it as the throne of kings.
They curse the time they cannot pass—
Then clip the angel's silver wings.

46

They sell their silence to the loud,
And call it wisdom not to speak.
They cheer the cruel, join the crowd—
Then hide when justice dares to seek.
They trust the mask more than the face,
And wear their pride like sainted fire.
They name deceit a form of grace—
Then punish those who still aspire.

47

They drown the spirit in a flask,
Then ask why life feels so unreal.
They build excuses mask by mask,

And say they've healed when they won't feel.

They beg for purpose, scorn the guide,

They scream for meaning, flee the cost.

They keep the truth just out of sight—

And call it found when it is lost.

48

They fill their shelves with hollow books,

And frame their walls with hollow art.

They practice empathy in looks,

But leave no room for broken heart.

They teach their youth to never need,

And call that strength, not silent fear.

They clothe their lack in branded creed—

Then cry when no one draws too near.

49

They burn their flags to light a stage,

Then dance atop the buried past.

They roar with all the scripted rage,

And hope their moment somehow lasts.

They bow to fame, then curse the lens,

They kiss the flame that scars their skin.

They claim the world in bitter ends—

Then wonder why they can't begin.

50

They touch the void, then call it fate,

They curse the rope, then climb it fast.

They swallow shame and call it plate,

Then say the hunger couldn't last.

They build their worth on what decays,

And pave their joy with soot and stone.

They run through brightly burning days—

Then sit in ash, and die alone.

51

They mark their skin with sacred signs,

But never let the symbols sink.

They build cathedrals from their spines,

But worship mirrors when they blink.

They pray in tones of irony,

Then curse the void that doesn't speak.

They brand their fears as destiny—

And sell the strong for looking weak.

52

They laugh at hunger from their thrones,

Then binge on fears they manufacture.

They write their laws in broken bones,

And make their prophets court the butcher.

They pardon crimes that bring them gain,

Then jail the saint who speaks of rest.

They weep for show, then toast the pain—

And leave the wounded dispossessed.

53

They train their hearts to never bleed,
Then wonder why the pulse is faint.
They auction faith for viral speed,
And craft their grief into complaint.
They feed the poor with hollow words,
Then feast behind cathedral walls.
They sing like tame, ironic birds—
Too proud to answer when God calls.

54

They crown the fool, then quote his song,
They scorn the wise who will not bend.
They claim the right to right the wrong,
Then wound the truth they can't defend.
They take the throne with vacant eyes,

And lead the blind in circles wide.

They say the fallen must arise—

Then laugh as they refuse to guide.

55

They fill their veins with bitter wine,

Then write epics of their despair.

They say the ache must be a sign—

Then drag their wounds from chair to chair.

They host their pain like honored guest,

They polish chains and kiss the key.

They say that peace must fail the test—

If peace does not come easily.

56

They pave their paths with shattered oaths,

Then wonder why their footing slips.

They speak of truth with poisoned throats,

Then hush the voice with numbing sips.

They write their myths in rented space,

And scroll through prophets they ignore.

They want the ending to embrace—

Without the journey to the door.

57

They seek redemption in applause,

And sell their sorrow in a jar.

They post their pride as hidden flaws,

And chase their shadow like a star.

They bear no weight but curse the wind,

And fake the tear, then fake the grin.

They say they never meant to sin—

Then build a shrine to where they've been.

58

They flee the dark, then curse the day,

They kill the flame, then fear the cold.

They make a game of wasting clay,

Then ask why nothing turns to gold.

They build their towers tall and thin,

Then jump and hope the ground forgives.

They cannot see the path within—

So they believe that no one lives.

59

They cover rot in shining stone,

And call the shell a work of grace.

They live together, die alone,

And fear the unfiltered face.

They beg for someone else to change,

Then dodge the mirror on the wall.

They say the world is growing strange—

But they have made it so, in all.

60

They line the streets with broken dreams,

And drive through wreckage dressed in style.

They say the end is what it seems,

Then race to meet it with a smile.

They buy their peace in desperate haste,

And poison it to make it last.

They name their future boundless waste—

Then mourn the shadow of their past.

61

They numb the soul to save the skin,

And teach the heart to play pretend.

They run from where the wounds begin,

Then call the break a holy end.

They chase the thrill of being lost,

Then blame the stars for every turn.

They warm their hands at any cost—

But never stay to feel them burn.

62

They mimic love, but not the cost,

And craft desire from idle flame.

They sell the truth that they have lost,

Then learn to smile without a name.

They change their face to match the crowd,

And curse the self that will not die.

They shout their secrets long and loud—

Then weep when no one asks them why.

63

They sculpt their grief into a brand,

Then wear it proudly on their chest.

They say they never meant to stand—

But still demand to be the best.

They hunger for the sacred sound,

Then drown it out with flashing light.

They build a tower on no ground,

And call collapse their final rite.

64

They love the mask more than the man,

And cherish form more than the fire.

They ask not whether they still can,

But only how to stoke desire.

They call the thirst a sign of soul,

Then never drink from living springs.

They praise control, condemn the whole—

Then wonder why the caged bird sings.

65

They polish pain until it shines,

Then sell it as a tale of grace.

They say that guilt comes from old lines—

Then draw new ones across the face.

They climb the mountain just to post,

Then flatten peaks to claim the sky.

They haunt their past like some proud ghost—

Afraid to ask the reason why.

66

They gut the word, then speak it loud,

And trade its meaning for applause.

They march alone into a crowd,

Then celebrate the fatal cause.

They drink from fountains dry and cracked,

And call the dust a cleansing rain.

They call their faith a coded act—

And wear their doubt like holy pain.

67

They leave their elders in the cold,
Then chase their stories on a screen.
They say they do not need the old—
But mimic gods they've never seen.
They cry for roots while cutting trees,
They speak of peace while feeding war.
They echo prayers on bended knees—
Then stand and leave through darker doors.

68

They sharpen love to make it sting,
And paint desire in bitter tones.
They forge their bliss from hollow things,
And fear the silence in their bones.
They stretch their joy until it snaps,
Then mourn it like it was betrayed.
They worship stars through broken apps—
Then wonder why they feel afraid.

69

They chase the end, then call it birth,
They break the wheel and call it free.
They mix their poison into mirth,
And sow their ruin carefully.
They promise light, then sell the smoke,
They teach despair as sacred rite.
They twist the vow into a joke—
And call the dark their guiding light.

70

They poison truth with urgent tone,
Then blame the lie for being told.
They cry they've never felt alone—
But never touched another soul.
They call themselves the rising dawn,
But only cast a deeper shade.

They shout until the voice is gone—
Then ask why songs no longer stayed.

71

They crown the wound and veil the scar,
Then worship what they've never healed.
They drive their faith into a jar,
And break it when the truth's revealed.
They chase the fire, then curse the flame,
And run from ashes they create.
They light the fuse, deny the blame—
Then call the end a twist of fate.

72

They break the dawn with sleepless eyes,
And trade the stars for plastic pearls.
They fill the night with distant cries,
Then drown the silence in new swirls.

They call it art, they call it free,

But never know what freedom takes.

They sell their souls in packs of three—

Then wonder why the spirit aches.

73

They wrap their greed in charity,

And wear false virtue like perfume.

They mock the prayer for clarity,

Then wonder why they choke on gloom.

They drain the meaning from each rite,

Then build new altars to their pride.

They ask the dark to grant them light—

Then hide from dawn when it arrives.

74

They cry for peace, then swing the sword,

And feast upon the war they start.

They speak in tongues that can't afford
The burden of a softened heart.
They label violence as defense,
Then march beneath a borrowed sky.
They forge their flags from burned incense—
And teach their children how to lie.

75

They grind their time into a paste
To smear across the void of day.
They run in circles, call it haste,
Then curse the hours that slip away.
They fill their minds with empty things,
Then grieve the loss of something real.
They chain themselves with golden rings—
And say they love the way they feel.

76

They flee the questions they once asked,
Then mock the ones who still believe.
They claim no purpose in their task,
Yet mourn each chance they cannot grieve.
They trade their wonder for a guide,
Then hate the path they're told to tread.
They plant their flags in shifting tide—
Then sink, and blame the sky instead.

77

They drink the storm, then sell the drought,
And cry when nothing flowers near.
They kiss the lie and spit it out,
Then blame the mirror for the smear.
They say their hearts are forged in flame,
Yet freeze when mercy dares to speak.
They chase the high, then curse the name
Of those who find what they still seek.

78

They beg the grave to give them time,

Then waste it writing in the dust.

They steal from prophets every rhyme—

Then say the world has grown unjust.

They stand where giants used to pray,

And mock the stillness of that place.

They carve their gods from cheap display,

And frame them in a user's face.

79

They sell the seeds, then scorch the field,

And call it progress when it dies.

They hoard what once was meant to yield,

Then drown their guilt in lullabies.

They draw a smile, then crack its shape,

And say the fracture makes it true.

They dress in skins they cannot fake,

Then ask the sky for something new.

80

They hold the past like it's a knife,

Yet wield it only on the weak.

They curse the price of sacred life,

Then charge admission just to speak.

They flood the earth with careless song,

Then mourn the silence they create.

They ask what's right, but follow wrong—

And praise their fall as sovereign fate.

81

They tax the soul for fleeting gain,

Then spend it all on empty thrills.

They medicate the sacred pain,

And numb the call that gently stills.

They buy the fire, not the heat,

They trade the truth for what deceives.

They leave the garden at their feet—

Then claim it vanished with the leaves.

82

They lift the thief, dismiss the wise,

And forge new morals out of ash.

They bottle tears to advertise,

Then light their fears and call it cash.

They kill the roots to cut the rot,

Then paint the stump and call it saved.

They curse the peace they once forgot—

Then beg to not be bound or braved.

83

They dress their hunger up in charm,

Then flirt with death in satin beds.

They mock the shepherd for the harm
That wolves have wrought upon their heads.
They wear their doubt as fine couture,
And pin their shame on shifting fads.
They ask for love to feel secure—
Then leave as soon as someone adds.

84

They broadcast every fleeting thought,
But bury every quiet grace.
They name the soul a thing ill-taught,
Then sell their spirit for a face.
They prize what glimmers, scorn what grows,
And pave their dreams with broken stone.
They mock the stream that gently flows—
Then thirst when left to drink alone.

85

They trade the slow for instant bliss,
Then curse the flash when it won't last.
They crown their wounds with poisoned kiss,
And haunt themselves with buried past.
They wear a smile with sharpened teeth,
And call the bite a kind of care.
They bury meaning just beneath
The surface of their vacant stare.

86

They legalize what once was shame,
Then call it cruel to not comply.
They change the words, then shift the blame,
And say the truth was just a lie.
They burn the maps, then curse the maze,
And say the path was made unfair.
They live in twilight, call it praise—
Then wonder why the dawn's not there.

87

They strike the match, then cry at flame,

And sue the heat for being warm.

They chant injustice with no name,

Then lead the mob to chase the storm.

They grieve what they themselves have done,

Then hire priests to mask the guilt.

They sell the shadows of the sun—

Then say it's joy they've always built.

88

They twist the meaning of the word,

Then cry when silence does not bless.

They cage the song of every bird,

Then mourn the lack of wilderness.

They say the heart is out of style,

They say the truth is just a phase.

They walk alone in single file—
And dream of unfamiliar praise.

89

They fear the fire that purifies,
Yet dance with flames that only burn.
They call each fall a new disguise,
Then build no bridge, but still return.
They cry for light, then draw the shade,
They plant the weed and curse the vine.
They worship stars already made—
Then ask to see a clearer sign.

90

They speak of hope as if it's dust,
Then mock the one who kneels to pray.
They watch the sacred slowly rust,
Then sell the rust to paint the day.

They hollow out the very word

That once could rouse the dead from sleep.

They name the soul a dying bird—

Yet wonder why they cannot weep.

91

They banish saints to praise their kings,

Then wonder why the throne feels cold.

They chain the bird but crave its wings,

Then weep when none will sing the old.

They mock the silence of the sage,

And call the desert cruel and bare.

They read the script, but skip the page

Where grace once asked them just to care.

92

They flood the void with restless sound,

Then tremble when the stillness speaks.

They crush the rose into the ground,

Then ask why beauty never peaks.

They strike the bell that won't return,

They break the hands that tried to heal.

They sell what they refuse to learn—

Then say the sacred isn't real.

93

They clothe their guilt in clever phrase,

Then sell the style to higher bids.

They quote the dead in empty praise,

Then hide their hearts like frightened kids.

They shout for change, but only move

In circles carved by comfort's chain.

They say there's nothing left to prove—

Then curse the sky for sending rain.

94

They scan the sky for signs of end,
But miss the roots beneath their feet.
They cannot tell a foe from friend,
But always know which mask to meet.
They ask for truth, then turn their back
When it appears in tattered cloak.
They light the fuse, then paint it black—
And swear the fire was just a joke.

95

They fear the cost of being known,
So build their walls with pixel thread.
They ache for something not their own,
Then fear the life they might have led.
They silence all who see too clear,
Then call the prophets mad with pride.
They sow with doubt, they reap in fear—
And call it strength the way they hide.

96

They stack their corpses in the cloud,

And code their sorrow into memes.

They fear the quiet, praise the loud,

Then haunt themselves in looped extremes.

They name their illness "self-control,"

And medicate the voice within.

They strip the silver from the soul—

Then say they never meant to sin.

97

They chant of rights with hollow breath,

But never speak of what is owed.

They joke of angels, laugh at death,

Then wonder why the road's corrode.

They seek no shepherd, shun the gate,

But cry when wolves begin to bite.

They curse the hour, question fate—
Then claim they've always known the night.

98

They carve their names into the dust,
And swear the wind will not erase.
They barter love for things that rust,
Then wear decay like saving grace.
They twist their sorrow into fame,
Then tell the young to do the same.
They mock the fire, play with flame—
Then cry when no one speaks their name.

99

They beg for time while wasting breath,
And name the waste a sacred rite.
They flirt with gods, then call it death
When none appear to make things right.

They scratch the veil, then call it torn,
They trace the stars but never kneel.
They claim the rose should come without thorn—
Then rage when beauty dares to feel.

100

They scream that all is meaningless,
Yet paint their cries in shades of gold.
They sell their lives for even less—
Then write the price as something bold.
They build their kingdoms, starve their kings,
And call their ruin majesty.
They fear the truth that mercy brings—
So die proclaiming: "I am free."

"Vanity of vanities, saith the Preacher... all is vanity."

— Ecclesiastes 1:2

ChatGPT said:

Updated saved memory

Thank you for your honest and powerful request. What you're asking for is no small thing — a 200-stanza autobiographical poem chronicling a life consumed by addiction, greed, distraction, and spiritual emptiness, with unflinching depth and truth. It will be raw, lyrical, and personal, drawn from the spirit of lament, confession, and bitter self-recognition.

I'll begin by drafting the first 20 stanzas, and proceed from there in consistent batches if you approve the voice and tone. The poem will remain grounded in your perspective — a life once devoted to false gods, now laid bare.

"The Feast That Made Me Hollow"

An Autobiographical Lament in 200 Stanzas

1

I fed my soul to neon flame,

And called it joy, and played the game.

I wore the mask, I knew the lines—

Each pleasure carved its claim in time.

I hunted more, I feared the still,

I chose the want, refused the will.

And every taste, and every lie—

I swallowed whole. I never cried.

2

The mirrors smiled when I would grin,

But never showed what lay within.

The boy I was grew thin and cold,

Beneath the weight of what I sold.

I bought the lie that I was king,

Then bowed to every lesser thing.

I claimed the night, denied the sun,

And lost myself before I'd begun.

3

I drank until the clock stood still,

Then drank to make it move again.

I broke my name on every pill,

And woke with strangers, void of friend.

I told myself I had control—

But watched my years slip down the drain.

I made a feast of ash and coal,

Then cursed the air for tasting pain.

4

I gambled hope in seedy halls,

With sweaty palms and shallow breath.
I kissed the dice, ignored the calls—
Of love, of truth, of looming death.
I said it's all just chance and fate,
But knew the dealer knew me well.
I dressed my losses up as great—
And laughed while slipping into hell.

5

I preached of freedom, mocked the bound,
While chaining every part of me.
I screamed for silence, begged for sound,
But fled from what might set me free.
I loved the edge, the fall, the flight—
The thrill that cost my soul too much.
And every time I missed the light,
I found another wound to touch.

6

I sought my worth in lust and flesh,

In fame, in fame's decaying breath.

I watched my soul grow less and less,

Until it barely shadowed death.

I touched, but never truly knew.

I kissed, but never felt the grace.

And when the morning cut me through,

I could not bear to see my face.

7

I mocked the sacred, praised the fake,

I worshipped idols shaped like me.

I lit the fire for ego's sake,

Then danced around it endlessly.

I scorned the meek, adored the loud,

I made myself the final god.

But every time I drew a crowd,

The silence underneath was broad.

8

I bought the best, I chased the new,

I bathed in envy's greedy tide.

I laughed at those who still were true—

Then cursed the love I had denied.

I mocked the ones who knelt to pray,

Then wondered why I felt no peace.

I gave my heart to games that slay—

And called the ache a masterpiece.

9

I overdosed on empty nights,

On screens, and sex, and swirling smoke.

I traded grace for flashing lights,

Then wore despair like some cruel joke.

I slept with fear beneath my bed,

I kissed regret and named it thrill.
I carved my soul until it bled—
Then blamed the world for standing still.

10

I chased the dragon, begged the flame,
I let the hunger write my laws.
I pawned my family, face, and name—
To feed a beast with sharper claws.
I bled for highs that left me low,
I sold my breath for fleeting fire.
And every time I said "hello,"
It sounded more like "I expire."

11

I wore my pain like suits of gold,
To prove I'd lived, to prove I'd won.
But underneath, my blood ran cold,

A shadow fleeing from the sun.

I told myself, “You are the storm.”

I never asked what winds had stirred.

And when the chaos lost its form,

I realized no one heard.

12

I burned my bridges, drank the smoke,

Then coughed and swore it tasted sweet.

I laughed as every promise broke,

Then wondered why I knew defeat.

I threw away the years like trash,

Each moment wrapped in shameful glee.

Then wondered why my soul would crash,

When all I’d built was mockery.

13

I glorified the endless take,

The empty chase, the constant fill.

I named each craving "holy ache,"

Then found myself both starved and still.

I stood in rooms with flashing walls,

But never found a single friend.

And every time the silence calls,

I hope it means my shame might end.

14

I wore the skins of lesser men,

To hide the rot beneath my chest.

I said "I'll start again, again,"

But never truly gave it rest.

I fed on vanity and pride,

On compliments and curated lies.

But even when I laughed and cried,

My soul was somewhere else—disguised.

15

I hollowed out the sacred grove,
To plant a tree of plastic leaves.
I traded wisdom for a stove
That only ever taught me grief.
I cooked my dreams in shallow pots,
And served them cold to those I hurt.
Then blamed the world for all my knots—
While sleeping drunk beneath the dirt.

16

I once had love. I once had faith.
But sold them both for something cheap.
A faster car, a smoother wraith,
A drunken night, a deeper sleep.
I found new gods in every face,
And kissed the devil with a grin.
Then looked for mercy, prayed for grace—

But could not name the place I'd been.

17

I fell asleep in stranger's arms,

But never once felt truly held.

I chased the high of surface charms,

While knowing something deeper failed.

I watched my children grow and fade,

I watched my lovers turn to stone.

I called it "living unafraid"—

But feared to ever be alone.

18

I learned to lie with elegance,

To say "I'm fine" through bloodied lips.

I toasted hell with opulence,

And danced on sinking pirate ships.

I made a show of what I'd lost,

I painted ruin like a prize.

But never once could count the cost—

Of all I'd burned behind my eyes.

19

I walked the streets with empty shoes,

And called it "freedom from the man."

I played to win, I played to lose—

But never once believed I can.

I filled my mind with broken glass,

And let it cut my every thought.

I called the future what must pass—

Then sold the past for what I bought.

20

Now here I sit, a vacant man,

With ghosts for friends and time to spare.

I used up all the life I ran—

And never stopped to really care.

I see the wreckage in the mirror,

I hear the stillness call my name.

And though the end is drawing nearer—

I can't say who it is I became.

21

The hunger never went away,

It only learned new masks to wear.

A sharper suit, a cleaner tray—

But still, the poison lingered there.

I called it taste, I called it class,

I called it "what I truly need."

But I was drinking broken glass,

And smiling as my gums would bleed.

22

I mocked the meek. I scorned the wise.
I cheered for men who built their thrones
On lies and sweat and lesser cries,
While silencing my father's bones.
I said the strong must take their due—
I thought the world was mine to own.
But now I see what grew askew,
And know how much I stood alone.

23

I turned my rage into a crown,
I ruled with sneers, with fists, with cold.
I crushed what dared to weigh me down,
Then wondered why no hand I'd hold.
I fed on wrath until it fed
On every remnant of my peace.
I dreamt of justice, painted red—
But never sought the soul's release.

24

The porn, the pills, the petty games,
The lights that blinked through every night—
They carved my body full of names,
But never once restored my sight.
I thought indulgence made me free,
I thought rebellion had a cause.
But all I did was bury me
In deeper graves with louder jaws.

25

I watched the good ones drift away—
Too calm, too kind, too clean for me.
I mocked their words, I turned to play,
I feared the still simplicity.
I said, “They’re weak,” but now I know
That strength is not the scream or sword.

It’s in the man who dares to grow—
Who walks away from what’s adored.

26

The gluttony was never food,
It was the need to always fill.
A hole that ate both pain and good,
And widened more with every thrill.
I licked the grease from silver trays,
I drank from cups that bled regret.
And though I smiled through all my days,
My heart was drowning in its debt.

27

I turned away from holy things,
And thought the stars were just old light.
I clipped the feathers off my wings,
And told myself, “This feels more right.”

No heaven watched, I used to say,
No purpose waits beyond this noise.
But now I ache to kneel and pray—
Not for reward, but to rejoice.

28

The gambling was not just the bet,
But gambling I had time to burn.
I threw away my youth's reset,
And never paused to truly learn.
I gambled trust, I gambled sleep,
I gambled love for passing thrills.
And all I earned was silence deep,
And years that echo down the hills.

29

I let the screens replace the sky,
The click replace the quiet tree.

And every pixel told a lie
That I mistook for clarity.
I measured joy by likes and shares,
I filtered every flaw and scar.
But in the dark, no one still cares—
You cannot friend who you are.

30

I made my body just a shell—
A means to hold the endless crave.
I gave it over, couldn't tell
What part of me was left to save.
I bruised it for another night,
I numbed it when the shame grew bold.
And when the mirror caught the light,
It showed a man already old.

31

I laughed at those who feared the grave—
As if I'd beat it with my charm.
But every drug, and every rave,
Just led me closer to its arm.
I made a joke of death and fate,
And played the fool with slick disdain.
But now I see that every hate
Was just a mask for hidden pain.

32

I shamed myself in subtle ways,
By calling weakness some new creed.
I made the worst of human days
My altar, and my broken feed.
I thought rebellion made me brave,
But never built a single path.
Instead I learned to how to cave,
And buried joy beneath my wrath.

33

My friends? They burned the same as me.
We passed the torch from hand to hand.
Each thought the fire meant we're free—
Not one could truly understand.
We laughed, we wept, we bled, we lied,
We crashed in piles of gleaming rot.
And now, in age, as they have died—
I wonder what the fire bought.

34

I thought forgiveness was a scam,
A fairytale for those afraid.
But now I beg the risen Lamb
To pardon what my hands have made.
The years I cast into the pit,
The love I broke, the words I killed.

If there is grace, then let me sit—
At least among the faintly willed.

35

The churches seemed too clean, too pure.
I didn't fit, so I walked past.
But filth became what I'd endure—
While swearing that it all would last.
Now I stand outside once more,
Afraid to enter, yet I stay.
Not for the wine, not for the lore—
But to unlearn the games I play.

36

I told myself I had no guilt—
That shame was just a myth for fools.
But now I see the life I built
Was framed in shadows, bound by rules.

Each sin, a thread in some grand net
That caught me slow, then pulled me in.
And every thrill I can't forget—
Is carved in bone beneath my skin.

37

I once had children, once had dreams.
But traded both for speed and smoke.
I blurred their cries with TV beams,
And missed each word they softly spoke.
I see their eyes now, far away—
And wish I could rewind the tape.
To say: "I'm sorry, please just stay,"
Before they learned how far I'd break.

38

I mocked the sober for their lack,
As if restraint was just a cage.

But I was bound, and they stood back—
While I collapsed from rage to rage.
Now I see what I once scorned
Was wisdom dressed in quiet clothes.
And every bruise that I adorned
Was chosen for the path I chose.

39

I had a wife. She tried to stay.
She cried for years, then walked at last.
I made her love a game to play—
Then cursed her when the spell had passed.
I said she changed, but that's a lie.
I changed, and never held her hand.
Now I just sit, and wonder why
I let go what I could not stand.

40

And so the feast goes cold at last.
The wine is gone. The lights are dim.
The ghosts of every reckless past
Now line the room and stare at him.
They do not speak. They only watch.
They do not judge. They only know.
And I, who once could never stop,
Can only now move very slow.

41

The money flowed like poisoned wine—
Enough to dull, not heal the ache.
I bought my face, I bought my spine,
And left no joy I didn't fake.
The richer I became in gold,
The poorer grew my inner land.
I owned the night, but not the hold
Of one true soul who'd take my hand.

42

I cursed the dawn, I feared the light.

I lived for neon, smoke, and screens.

I fed the beast of every night,

And sold myself in shattered scenes.

The man I was began to rot,

Replaced by masks I couldn't name.

Each one denied the me I'm not—

Each one disguised the deeper shame.

43

The mirror cracked, but still I looked,

Hoping to find a truer face.

But all I saw were years I took,

And stole from love, and time, and grace.

I practiced smirks, I rehearsed charm,

But nothing masked the hollow man.

I raised the gun, I raised the harm—
Yet couldn't raise a single plan.

44

I filled the silence with a song
Composed of lust and bitter pride.
The verses said that I belonged,
But in the chorus, truth had died.
I played it loud. I danced. I laughed.
The melody disguised the scream.
But when the needle snapped in half,
I heard the ending of the dream.

45

I called my mother once a year,
Too drunk to speak of love or pain.
I left my father cold with fear—
He knew I'd never change again.

They waited, prayed, and watched the skies.

They asked for signs, and lit a flame.

But I had traded all their cries

For blinding heat and hollow fame.

46

My body begged for rest and air.

It whispered, "Please, just let me be."

But I had demons left to dare—

And sought no space to set them free.

I fed it liquor, smoke, and lies.

I told it, "This is how we win."

Now all it does is shrink and sigh

Inside this stretched and withered skin.

47

I mocked the past as weak and blind,

As if new sin made me divine.

But every future that I mined
Was just a grave I made look fine.
I told myself, "This is the way—
The strong devour, the bold enjoy."
But now the strong are lumps of clay,
And I'm an old, discarded toy.

48

I met a girl who saw me break.
She stayed for months—then ran for good.
She tried to help, to mend, to wake—
But never knew just where I stood.
She said, "You're lost inside a maze
That you designed to trap your soul."
And I just lit another blaze
To chase away her quiet goal.

49

The sex was endless, hard, and bare—
A race to nowhere, through the void.
I never felt the body care,
Just used the flesh to feel destroyed.
I used my charm to draw them close,
Then used my pain to push away.
And every touch I feared the most
Would echo what I'd never say.

50

I used to smile at other men
Who built their homes and raised their kids.
I said, "They'll rot like me—but then,
They'll never live like I just did."
But now their lives look clean and whole,
While I just count what I have lost.
I chose the fire, not the coal—
I chose the thrill, and not the cost.

51

I filled my walls with screens and noise.

I banned the books. I trashed the still.

I said the world was made for boys

Who chase their want and take their fill.

But when the battery ran dry,

And all the lights began to fade,

I found no stars left in the sky—

No voice, no vow, no holy shade.

52

I laughed at men who bowed to God.

I said, "Why kneel? Why speak to air?"

But now I feel my heartbeat throb

Like one last knock on death's cold stair.

I do not beg, I do not plead—

But I regret the years I mocked.

For now I see how deep the need

To ask for what my heart once blocked.

53

The parties blurred into the years.

The faces changed, the laughter died.

I danced through failure, drowned my fears,

And buried every friend who tried.

I told myself, "This pain will pass,"

As if time ever fixed the wound.

But I just drank a deeper glass

And prayed to sleep inside the tune.

54

The jobs were shallow, sharp, and fast.

I chased the next, then dropped the last.

I couldn't care, I couldn't stay—

I only lived from day to day.

The résumé became a scroll
Of conquests faked and bridges burned.
Now I have nothing but a hole
Where once ambition faintly yearned.

55

I wore designer sins with pride.
I quoted Freud, I referenced pain.
I told the world I'd never hide—
Then ducked inside the same refrain.
I smoked my mind, I dulled my edge.
I climbed atop a house of shame.
I swore I lived without a pledge—
Then choked when someone said my name.

56

There was a night I nearly died.
I saw the edge. I heard the dark.

The needle slipped, my chest grew wide,
And silence took me like a spark.
But death refused to let me go—
It tossed me back like chewed-up meat.
I woke in vomit, slow and low—
Ashamed to stand on sober feet.

57

I laughed the next day like a joke,
As if survival meant I won.
But part of me stayed curled and broke—
The fearless child was now undone.
And though I drowned that voice again,
It always murmured through the haze:
"You've wasted love, you've murdered men—
And time won't give you back your days."

58

I see the ruins that I built
From pride and shame and shifting sand.
A crooked throne, a tower of guilt,
No friend, no guide, no steady hand.
I thought I'd rule a world of fire,
But all I did was torch the path.
I fed my soul into the pyre,
And called that funeral my wrath.

59

I see now, too, the lives I touched—
Not gently, but with hunger's grip.
The ones who trusted me too much
Now vanish like a sinking ship.
They tried to love, I took, then ran.
I used their hope, I drained their will.
I made a weapon of the man
I forged from lust and broken thrill.

60

And yet—I breathe. The morning comes.
My bones still ache, my hands still move.
I sit amid the ash and crumbs,
And wonder what I still could prove.
I am not young, and time is thin.
But still, the wind outside is sweet.
Perhaps, if not forgiven—then
At least I'll stand on cleaner feet.

61

I tried to write a single prayer,
But found my words were made of smoke.
They curled and vanished in the air,
As if the voice of truth had broke.
What gods would listen now to me—
A man who danced on every creed?

I’ve worshipped pride. I paid the fee.
I sold my mercy for my need.

62

My friends were ghosts before they died.
We shared a pact to chase the void.
We laughed when better men had cried,
And crushed what others had enjoyed.
Now I can’t name the ones I lost—
Their names erased by pills and gin.
I wonder if they count the cost
From far below or deep within.

63

The lovers blurred into the bed.
I never learned their final names.
We parted with no tears or thread,
Just echoes scorched in pleasure’s flames.

I told them lies to keep them near.

I kissed their dreams, then turned to dust.

And now, when someone draws too near,

I see in them the death of trust.

64

My mind is cracked but still it holds

A ledger of the things I've done.

Each entry speaks in iron folds

Of moments lost and races run.

There was a child I never met,

Because I fled the morning call.

She might be grown. She might regret.

I only know I dropped it all.

65

I mocked the sacred and the still.

I scoffed at monks and men of peace.

I said that truth was just a pill

For those too weak to seek release.

But now the silence haunts my ears.

It speaks in sighs I can’t ignore.

I traded awe for petty sneers—

And stand now lost outside the door.

66

I fed the lion of desire

Until it turned and ate my hand.

I warmed myself beside the fire,

Then cursed the heat I couldn't stand.

No hunger sated, thirst still burns.

No joy endures beyond the night.

Each thrill I grasped, the morning turns

To guilt that bleaches out the light.

67

The pills, the lines, the acid tabs—
I chased the edge beyond the edge.
I laughed as sense broke into jabs,
Then staggered off the rooftop's ledge.
I didn't fall—but wish I had.
The landing might have meant an end.
But every breath just made me mad,
And every fix betrayed a friend.

68

The gambler's heart beat in my chest.
I bet on ruin, doubled down.
I called my debts a holy test,
Then drowned again in shame's old gown.
The wheel would spin, the dice would roll,
I praised the chance as if divine.
But always lost a piece, a soul—
And never claimed what might be mine.

69

I paid for joy, I leased delight.

I bartered years for instant high.

I lived as if each neon night

Could keep the morning from the sky.

But dawn was cruel and never late.

It dragged my truths into its glare.

And in that light I met my fate—

A hollow man, a cold, blank stare.

70

The whiskey laughed when I could not.

It sang me songs of gentle hate.

Each bottle sealed another plot

To toast regret and call it fate.

I kissed the rim, I drained the glass.

I shook with sweat. I begged for more.

But joy would always fail to last—

And shame would vomit on the floor.

71

I made my bed in clubs and bars,

Where strangers touched and never stayed.

Where lights would mimic fallen stars

And every vow was soon betrayed.

The music throbbed, the bodies swayed,

The hunger pooled like heated wine.

And though I grinned, and though I played,

I never felt the dance was mine.

72

I heard a preacher once, outside—

A streetlamp halo on his head.

He called to souls who tried to hide,

And wept for hearts already dead.

He looked at me and didn't speak.

His eyes were full of something vast.

I felt my armor start to leak—

Then turned away and walked on past.

73

The years collapsed like empty shells.

They cracked beneath my swollen pride.

Each echo rang like funeral bells

That marked the love I pushed aside.

I do not mourn the things I bought.

I do not crave the fame I earned.

But every time a child is taught,

I weep for what I never learned.

74

I chased the art, but not the craft.

I scribbled noise and sold it bold.

They praised the style, they cheered the shaft—

But not the wound that should be told.

I made my name a brand to sell,

Then sold the brand for quick applause.

Now no one reads the tales I tell—

And silence is my last applause.

75

I made a temple of my skin,

And carved new idols in the flesh.

The ink was loud, the pain wore thin,

But none of it could make me fresh.

I pierced and marked to feel alive,

To own the frame I'd once ignored.

But nothing made my heart revive—

It stayed the tomb I had adored.

76

The shows, the screens, the constant scroll—
I fed my eyes and starved my soul.
I let distraction take control,
And played the fool in every role.
I knew the actors, learned the lies,
But missed the scene where grace appeared.
Now every moment that flies by
Feels like the life I should have steered.

77

I mocked the men who lived for good,
Who knelt and served and built with care.
I called their kindness weakness, stood
Above them in my poisoned air.
But now, old hands and brittle knees,
I envy all they chose to be.
They raised their homes in faithful peace—
And I raised towers none can see.

78

I've tasted every kind of sweet,

And turned to salt in every taste.

I let my hunger write defeat

On every love I laid to waste.

The cake, the flesh, the drunken feast—

I gorged and gorged and gorged again.

And now, where once I danced as beast,

I curl like ash, afraid of men.

79

I broke the rules, I mocked the law.

I claimed that freedom had no line.

But found beneath the wildest awe

A deeper cage, a worse design.

For I was never truly free—

Just ruled by cravings, raw and blind.

And every “yes” that tasted glee

Would choke me with the chains behind.

80

Now silence sits beside my bed.

She watches while I cough and fade.

She whispers things I might have said

If I had not betrayed the shade.

But I am here, and I still breathe.

A life remains, though trimmed and scarred.

Perhaps a seed beneath this sheath

Still waits to bloom within the shard.

81

I never planned to be this old.

I thought I’d burn out in my prime.

The young believe they can't be told

What lives in hunger past its time.

But here I am, with cracking joints,
And trembling hands that once held fire.
Now every breath just sadly points
Toward ashes dressed in old desire.

82

The mirror mocks me with its truth.
The sagging jowls, the sunken eyes—
All carved by decades spent uncouth,
A history of alibis.
I used to smirk with lips uncracked,
Now every smile's a paper tear.
And every truth I never backed
Returns to haunt me unaware.

83

The memories blur, the names decay.
The stories blend, the dates collapse.

I once could charm the night to stay—
Now sleep is stitched with silent traps.
I dream of women I defiled,
Of friendships starved, of unborn sons.
I wake with sweat and smell the wild
Regret that drips when morning comes.

84

I gambled love for passing thrill,
And lost it all in borrowed light.
I claimed the pain, refused the will
To walk away or make things right.
She called me once, then not again.
Her voice still lingers in my chest.
I picture her with other men—
And know she made the wiser rest.

85

I heard my brother's heart give out.

He died alone, the TV on.

No angels came, no holy shout,

Just static where the man had gone.

I hadn't called in seven years.

He'd asked me once to make things new.

But pride had deafened both my ears—

Now all I have is grief untrue.

86

The church stood open down the block,

Its doors like arms I never knew.

I walked past slowly, let it mock

My emptiness and shadowed view.

I used to spit at cross and creed,

Call prayer a crutch for lesser men.

But now, with every unmet need,

I wish I could believe again.

87

I bought a book on sin and soul,

But left it sealed upon the shelf.

I feared the weight, the final toll—

Afraid to truly meet myself.

Instead I thumb through shallow things:

Old magazines and catalogues.

The Word awaits with burning wings,

But I just mumble with the dogs.

88

My bones now ache from old excess.

The joints are rust, the breath is thin.

The doctor says the tests suggest

A slow decay has bloomed within.

He asks if I have next of kin—

I laugh and lie and change the tone.

No one will hold me when it ends.

I built a kingdom all alone.

89

I still recall the parties' glow:

The lights, the laughter, glutted pride.

But in each face I now bestow

A ghost of what I tried to hide.

They saw me rich, they called me king—

But none could see my crumbling crown.

For every feast and diamond ring

Just weighed my tarnished spirit down.

90

The mansion echoes when I speak.

Its walls don't answer anymore.

The paintings seem to softly leak,

The rugs wear trails across the floor.

The silence here is not serene.

It scratches like a hungry rat.

The ghosts I've made remain unseen—

But feed on me, and grow from that.

91

I see a child across the street.

She waves at cars and spins in play.

Her joy is raw, her soul complete,

Unburdened by the games I'd pay.

I want to shout, to warn her loud:

Don't chase the things that rot and fade!

But I just sit and watch the cloud

Of who I was drift into shade.

92

I bought a gold-plated guitar,

But never learned to make it sing.

It hangs unused, a bright memoir
Of dreams I drowned for other things.
I'd trade ten thousand nights of vice
For just one song played true and right.
But skill, like love, demands a price
I always ducked to chase the night.

93

A letter came from someone dear—
A son I never thought to claim.
He said he found me, year by year,
And knew his silence bore my name.
He'd made a life beyond my reach,
Forgave me without knowing why.
His words were soft, a kind of preach—
And stabbed me harder than goodbye.

94

I held the paper to my face.

It smelled of ink and second chance.

I saw within his lines a grace

That danced like flame beyond my stance.

But guilt has calloused every nerve.

It teaches joy to ache with shame.

And though I read the love he served,

I still can't call him by his name.

95

I bought a burial plot last spring.

It overlooks a dying tree.

No one will come, no choir will sing,

But I chose it for privacy.

The clerk just nodded, took my check.

She didn't ask what I had done.

And that, I think, is how regret

Becomes a life forever run.

96

Sometimes I think of simple meals—

A bowl of soup, a slice of bread.

Not tasting just to chase new thrills,

But grateful simply to be fed.

I used to eat as if the act

Could drown the void inside my chest.

Now hunger shows me what I lacked:

A heart that knew when it was blessed.

97

The holidays bring scents I miss:

Of cinnamon, and pine, and light.

But all I know of joy is this—

A faded shape in black and white.

The children sing beneath my street.

Their voices lift, and I draw near.

Then hide, ashamed to let them meet
The man who choked on Christmas cheer.

98

I miss the smell of honest work—
Of sweat and sun, of hands well-spent.
I miss the ache I used to shirk
When labor gave the body vent.
I built no fence, I dug no trench.
I paid for hands to lift my load.
Now all that ease begins to clench
Around my throat like debt long owed.

99

I lit a candle late last night.
The wick took flame, then calmed the room.
It danced, not wild, but pure and bright—
A little sun against the gloom.

I watched it burn and thought of faith,
Of how belief begins so small.
And how, despite the wrecks I've made,
A spark might still outlive the fall.

100

So here I pause, but not to end.
The tale rolls on—my breath is proof.
If I have time, then I can bend
What's left toward something nearly truth.
Not clean, not whole, but maybe real—
A broken man who dares to try.
To walk with limp, to speak and feel,
And raise a prayer before I die.

101

They say confession sets you free,

But I've confessed to empty chairs.
I've told my sins to dead TV,
To walls that blink with coded stares.
I've typed my guilt on glowing glass,
And posted grief like some lost coin.
But none of that can cleanse the past,
Or stitch the soul that won't rejoin.

102

I once believed in no beyond—
Just meat and moment, flash and flame.
But now, with every breath less fond,
I ache to call the silence Name.
I crave a Judge, a Holy Eye,
If only so I might be seen.
To die in shadow terrifies—
To vanish, unloved, and unclean.

103

I tried to drown the voice inside—

The whisper that would not relent.

I cursed it, mocked it, watched it hide—

Then begged for it when strength was spent.

It spoke of kindness I betrayed,

Of vows I snapped like brittle thread.

It waited while I disobeyed,

And wept for every word I said.

104

A stranger left a note one day

Tucked gently near my mailbox slot.

It simply read: You're not the clay.

Then vanished, nameless, like a thought.

I pondered it for hours on end.

It made no sense, then far too much.

A message, maybe—sign from friend,

Or warning from a softer touch.

105

The liquor store is boarded now.

The lights went out last winter's frost.

I used to kneel beneath its vow

And pay with all the joy I'd lost.

I pass it like a grave, half-known—

A shrine to needs that never fed.

The ghost of who I was has grown,

But nothing it consumed is dead.

106

I've seen old footage of myself—

A wedding toast, a speech, a smile.

And all of it feels like a shelf

That never bore a thing worthwhile.

The suit is sharp, the teeth are bright.

The crowd applauds, the glass is raised.

But no one knew the silent blight—

The man beneath, unloved, unfazed.

107

The credit cards, the endless debts,

The watches I could never wear—

Each token marks the same regrets

That luxury could never spare.

I slept in sheets that cost a home,

I bathed in tubs too wide for one.

And all the while, I'd drift and roam—

A prince exiled from morning sun.

108

The drugs gave colors to my eyes—

A thousand lights the world denied.

But every high became a lie

That left me thinner when it died.
I chased the serpent through the maze,
I kissed the needle like a vow.
And now, in slow descending haze,
I nurse the scars upon my brow.

109

I mocked the men who worked for bread—
Called them blind slaves to simpler ways.
But now I envy where they tread—
Their quiet truths, their grounded days.
They built with sweat and ended poor.
I soared, and crashed with gilded wings.
Now every beggar at my door
Knows more of peace than I of kings.

110

I used to think the world was mine—

That nothing sacred lay above.
But I mistook the taste of wine
For anything that felt like love.
I thought the flesh could make me whole.
I thought the touch would calm the storm.
But every body that I stole
Just left me colder than the norm.

111

I heard a preacher once in rain.
He didn't shout, just calmly spoke.
He mentioned joy, and grief, and pain—
Then closed his eyes beneath his cloak.
I didn't hear a word he said,
Too drunk to follow half the tale.
But now his silence fills my head—
A voice I failed, a whispered grail.

112

The devil doesn't wear a mask—
He shows you mirrors, polished smooth.
He doesn't give you hell—he asks
What hell you'd build to soothe your truth.
He brings no fire, no iron chains,
No sulfur, pitchfork, blood, or drum.
He offers credit, pills for pain—
And cheers when you forget to come.

113

I touched a child's hand at the store.
Not mine—just one who looked at me.
He smiled, then turned, and asked no more—
A passing grace, a brief decree.
It felt like rain upon my skin.
It felt like something near divine.
And for a breath, I felt within

A hope that something might still shine.

114

I found an old, unopened gift

My mother sent before she died.

I couldn't bring myself to lift

The paper where her hands once tried.

But then I tore it, slow and dumb—

A scarf she knit, still soft and red.

I wore it once, then felt so numb

I wept beside her ghost instead.

115

The past does not return in flames—

It drips like water on a cave.

Each drop inscribes forgotten names

That echo in the holes we crave.

And in those hollows, something stirs—

Not rage, not guilt, not even pride.
Just shadows speaking quiet slurs
Of who I was before I died.

116

My father taught me how to cheat.
He said the world rewards the bold.
He showed me how to lie and eat
From plates that others hadn't sold.
He called it strength, and I believed.
He died with diamonds on his chest.
But now I wonder what we thieved
That might have bought a little rest.

117

The funeral came late one fall.
I wore a suit two sizes big.
No one from town showed up at all—

Just me, the casket, and the fig
Tree he had planted long ago.
It dropped one fruit upon the stone.
And in that fruit, I seemed to know
How little love he'd ever grown.

118

I stare at shelves of books unread.
Their spines uncracked, their knowledge sealed.
I bought them just to feel well-bred,
But never let their truths be healed.
The prophets speak of wrath and grace.
The poets mourn what men forget.
And all of them, in one long face,
Stare back at me with wise regret.

119

The mail comes late. The bills still cling.

The catalogs all beg and boast.
But nothing now feels worth the ring
Except the letters from a ghost.
A friend I wronged once writes each year,
Though I have never written back.
His grace persists, unbent by fear—
A thread of light along my track.

120

So here I sit, not dead, not healed—
But somehow open, cracked, and bare.
The wounds I spent a life concealed
Now breathe like lungs that need the air.
If I must die, let it be slow,
So every poison leaves its trace.
And maybe in that purge I'll know
The shape of mercy's hidden face.

121

There were no monsters in my tale,
No tyrants, demons, beasts to slay.
Just hungers dressed in decent mail,
And comforts that erode the day.
The couch became my iron throne.
The screen—my altar, cold and still.
And though I sat for years alone,
It always answered with a thrill.

122

I saw my neighbor die alone—
A quiet man with little name.
They found him months beyond the stone,
His dog still waiting just the same.
No wealth, no will, no final feast.
His world had simply slipped away.

Yet something in his quiet ceased
To haunt me more than what he'd say.

123

For I have spoken grand, loud lies
And heard the cheers my falsehood drew.
I fed the crowd with sugared sighs—
They clapped, and none of it was true.
And afterward, I'd drink alone,
Ashamed of words I'd finely trimmed.
A beast can thrive where truth has flown—
I made my peace with what I dimmed.

124

The hunger I have known is deep.
Not just for food, or touch, or gold.
It roots in marrow, claws in sleep,
And grows more fierce the more I hold.

I thought consumption made me large,

A god of taste, of will, of might.

But all the while, I was at charge—

A slave to more, a worm to bite.

125

I gambled once and lost a house,

Another time, a month's full wage.

But still I sat there, still a louse,

Still sure the next hand flipped the page.

It's not the cards that strip your soul,

It's what they promise you'll become.

A better man, a lucky goal,

A future humming like a drum.

126

My wife once begged me to come home—

Her voice a trembling autumn leaf.

But I had found a louder dome—
A barroom thick with numb belief.
She left, and took the children too.
I didn't follow, didn't ask.
I lit a smoke and claimed I knew
That family was just a mask.

127

Years later, at my daughter's door,
I knocked with hands too dry for grace.
She answered with a look I wore
When staring into my own face.
No words. Just silence in the air.
She didn't scream, or cry, or bless.
She simply turned, and left me there—
A ghost no grief could now address.

128

And still I chased the evening's sin,
Like night was something I could own.
The women came with perfumed skin,
But none could touch me to the bone.
Their eyes were tired. Mine were worse.
We'd speak in half-truths, pay, and part.
I wasn't seeking love or nurse—
Just something colder than the heart.

129

The years pile up like unread mail.
Each one unopened, thick with guilt.
You swear you'll face them without fail,
Then build another floor you tilt.
The house of self—so vast, ornate.
So many rooms I never saw.
But every window frames the fate
Of men who thought their want was law.

130

My body groans beneath its sins.

The liver aches. The stomach tears.

The tongue remembers too much gin.

The teeth recall the candy snares.

But worse than what the doctor names

Are things no test can diagnose—

The ache that says, through softer shames,

You never fed what mattered most.

131

I met a man with prison ink—

A cross etched deep into his skin.

He said, “It’s realer than you think—

The blood, the shame, the cell within.”

He’d done ten years for something foul.

He'd lost a brother, mother, son.

But in his voice, no hint of scowl—
Just grief, and work, and what he'd done.

132

He asked me what I'd ever built.
I said, "Regret." He said, "That counts."
He handed me a flask of guilt—
Sipped water while I measured mounts.
He said, "We're all just trying slow
To fix the fires we helped ignite.
But some of us don't even know
What it might mean to make things right."

133

I never prayed until the pain
Turned louder than the voice of pride.
Then all at once, a thin, cracked chain
Of syllables came up and cried.

I didn't ask for wealth or peace.

I didn't beg for some new start.

Just silence. Mercy. One release

From being torn apart by heart.

134

And though no thunder shook the sky,

No angel came, no vision burned,

I felt the stillness pass me by—

A moment that the soul discerned.

Perhaps that was the start of grace,

Or just the pause before the fall.

But something shifted in that place—

The first thing not consumed at all.

135

I saw a boy with nothing much—

Just ragged shoes and crooked grin.

He offered me his crust and such,

As if I were the starving kin.

I took it, stunned, then couldn't eat.

The hunger passed, replaced by shame.

For once, I saw through my deceit—

And knew the face beneath my name.

136

I watched a dying tree one day—

Its limbs like claws, its bark all split.

No leaves, no fruit, no shade, no sway,

Just something standing forfeit.

And yet it held one bird's small nest.

One fragile egg within its keep.

I thought: Perhaps what's cursed and stressed

Still guards the living in its sleep.

137

There's beauty in the broken things—
A cup once cracked, now used with care.
A song a hoarse old woman sings
That lifts despite the thinning air.
But none of this was mine to hold
Until I saw what I had done—
Until I wept, alone and old,
And stopped pretending I had won.

138

The television rarely speaks.
Its voice has dulled, its pulse has waned.
The shows repeat. The script now creaks.
The drama ends where none complained.
I shut it off and heard the rain—
Not metaphor, just falling light.
And somehow in that wet refrain
I felt the edge of something right.

139

I don’t yet claim to know what’s true.

I’ve worshipped idols, worn their chains.

But I have seen what lust can do,

And walked the halls of pleasure’s stains.

If there’s a God, He’s not like me—

Not starved for praise, not quick to strike.

He waits, perhaps, where none yet see—

Among the poor, the shamed, the like.

140

So now I write this broken hymn,

Not for applause or grand reward.

But for the voice that once grew dim

To know it’s heard, if not restored.

Two hundred stanzas may not cleanse,

But every word I lay in dust

Might meet the silence as a friend—
Or crack the coffin of my trust.

141

The world I built was made of noise—
Not music, mind you, not a song,
But groans from broken, rusted toys
That failed to hold me very long.
A flashing screen, a fast red car,
A louder laugh to hide the crack—
But always just beneath, a scar
That none of these could plaster back.

142

I used to mock the quiet ones—
The slow, the soft, the kind, the meek.
I'd sharpen tongues like loaded guns
And fire them at the mild and weak.

Yet in the end, I envy most
The gentle hearts I once dismissed.
They haunt me now, like some old ghost—
Each one I hurt, each time I hissed.

143

There was a friend I called too late—
A brother, really, from my youth.
I wore the years like armor plate,
He wore them like a truth.
He said he'd beat the drink, the pills.
He found a life beneath the ash.
I told him lies, he paid my bills.
Then died before I paid him back.

144

Debt's not just money, not just time.
It's every look I turned away.

It’s every love I made a crime,
Then cursed the world when none would stay.
It’s children taught by what I praised—
Quick pleasures, loud and consequence-free.
It’s sermons burned, it’s idols raised
To every warped version of me.

145

I wandered through a Sunday fair—
Old, limping, clutching cotton bread.
A boy ran past with windblown hair,
Alive in ways I thought were dead.
He spun, he laughed, he had no shame.
He saw me—waved without a care.
He didn’t know my face or name—
Just saw an old man standing there.

146

That wave cut deeper than a blade.

It said: You're still here. So begin.

It said the cost of all I'd made

Was not the end, just discipline.

I sat down on a rusted bench,

The sun a fire without a cage,

And felt—for once—the world unclench

Its fingers from my throat of rage.

147

The mirror was my last old friend.

It never lied but never healed.

It showed me how beginnings bend

When all you've sown remains concealed.

My jowls were heavy, eyes gone dim.

The flesh betrayed what soul had done.

But still, within the gray, the grim—

A flicker like a setting sun.

148

I thought I'd earn some final prize—

A medal for enduring pain.

But no one pins the selfish wise,

Nor builds a monument from shame.

The best I'll get is what I give:

A word to someone still in fire.

A reason yet for them to live—

A call to rise from old desire.

149

I saw a preacher in the park—

A folding chair, a crooked smile.

He talked of light and bitter dark

With words too plain to be in style.

He said the soul can rot from wealth,

That riches breed a certain death.

And though he coughed and lacked for health,
Each word he spoke gave me a breath.

150

I didn't kneel. I didn't cry.
No blinding flash, no choral sound.
But something flickered in the sky—
A bird that circled, then came down.
It picked at crumbs beside my shoe.
I watched it hop, then lift and dart.
And in that pointless, passing view,
I felt the beat within my heart.

151

The heart I buried years ago—
Beneath a feast of meat and wine—
Now thumped as if it longed to know
If I could still be something fine.

Not perfect, no. That ship had burned.
But maybe honest, maybe true.
And from that thought, some hunger turned
From what destroys to what renews.

152

I passed a shop where once I'd bought
A dozen things I never used.
The shelves now bare, the owner caught
Behind a counter, worn and bruised.
He said, "It's ending. Times are thin."
I nodded, lost for what to say.
We shared a silence like old kin
Who'd gambled most their lives away.

153

I walked past bars I used to haunt—
Each door a memory, each booth

A scar I fed with lies so gaunt
They almost passed for truth.
No bouncer stopped me. No one cheered.
Just empty stools and sticky floors.
As if the ghosts that once appeared
Had moved on to some brighter wars.

154

The wars we wage are seldom loud—
Most fought within, in mute decay.
Not with a sword or surging crowd,
But when we throw our soul away.
And I had tossed mine in with glee,
Sold piece by piece for fleeting light.
But now I see what used to be—
A man who feared the honest night.

155

I feared the dark because it told
The truth that silence could not bend.
It whispered, when the flesh was bold,
That all I built would meet its end.
And yet, it also gave me space—
No music, wine, or flashing flame.
Just air and shadow, death and grace,
And me at last without a name.

156

It's hard to start when age is near—
When knees are weak, when hands can't write.
But harder still to die in fear
That all you did was chase delight.
So here I am with pen in hand—
Not saint, not sage, not hero cast.
Just one old man who understands
How fast the feast becomes the fast.

157

No child weeps for what I was.

No friend remains to say my name.

I burned them with such strong applause

They never guessed it hid my shame.

Yet even now, some morning breaks.

A breath. A thought. A prayer half-formed.

And in that crack, the body wakes

To something faintly, strangely warmed.

158

Redemption isn't built like stone.

It doesn't rise on laws or vows.

It moves through marrow, breath, and bone—

A seed within a broken house.

I do not know if I'll be saved,

Or if such grace was meant for me.

But I will dig, though long depraved,
To find what roots still seek to be.

159

For even weeds can bloom in dirt,
And even ash can feed the rose.
The soul, though choked by self and hurt,
Still stirs beneath what darkness chose.
I do not ask for wealth or youth.
I do not ask to be made pure.
Just strength to face what was the truth,
And light enough to still endure.

160

So here I end—at least for now.
One chapter more, and still not done.
I do not seek to make a vow,
Just hold the line from which I run.

I have consumed and I've decayed.

I've fed on rot, I've begged for flame.

And now I write what I betrayed—

And dare at last to name my shame.

161

I tried to build my worth in steel,

In marble floors and polished stone,

But found no comfort I could feel

When all I had was mine alone.

The riches did not hold the chill,

Nor did the rooms I made expand.

What's wealth, when silence wraps you still

And no one's left to take your hand?

162

I bought applause with clever lies.

I paid with favors, charm, and wit.

But truth grew heavy in disguise—
A debt I carried bit by bit.
I fooled them all, but not the soul
That waited patient in the core.
I played the part, I filled the role—
But died a little more and more.

163

There were no dragons, no great foe—
Just hunger dressed in worldly clothes.
Just days that passed too fast, too slow,
With plastic crowns and plastic thrones.
My kingdom cracked with every thrill.
My laughter rang with distant ache.
And all I touched, and bent to will,
Would crumble faster than I'd make.

164

A woman once had loved me well—

She saw through postures, saw the man.

But I mistook her grace for spell,

Then broke her heart because I can.

I've dreamed of her in fractured ways—

In kitchens warm, in silence deep.

But nothing mends the blinding days

You chase the flesh and lose the sleep.

165

I squandered chances like spare change.

I tossed away the holy rest.

I worshiped every passing strange,

And scoffed at what was truly blessed.

The sacred seemed too slow, too still—

A thing for fools and dying men.

But now I beg to drink that will

That quenches fire without end.

166

There is no clock to turn around.

No hour returns once it is lost.

Each step I took on shifting ground

Came at a hidden, mounting cost.

I do not seek to buy it back—

The years I stole, the grace I fled.

I only seek to walk this track

With eyes that see, not sleep instead.

167

My hands once crushed the fragile thing—

A bird, a hope, a whispered plea.

I mocked the ones who tried to sing

Of better lives, of being free.

But now I envy every tone

That breaks the stillness with a hymn.

For silence is the tyrant's throne,

And I have sat too long with him.

168

The gamblers spoke in glowing tones

Of luck and lightning, fate and thrill.

But all their words were gnawed by bones

Of what they sacrificed for will.

I joined their games with poisoned pride—

And smiled while losing time and soul.

For none can win who's died inside,

And I was dead before the roll.

169

I searched in mirrors for a god—

A shape to mimic, face to prize.

But found instead a slowly flawed

Illusion dressed in aging guise.

I prayed once to my own success—

A golden calf with teeth and charm.

But idols never do confess

That all they bring is greater harm.

170

There was a night I cursed the stars—

Too bright, too far, too pure, too clean.

Their order mocked my tangled scars,

Their silence made my hungers mean.

But even then, a voice inside

Spoke not in rage but aching grace:

"You still are here. You have not died.

You still may turn and seek my face."

171

That voice returns at strangest hours—

A faucet drip, a furnace sigh.

Not grand like thunder, clouds, or towers—
Just breath against a darkened sky.
It asks for nothing but a pause,
A crack in armor, bare and real.
It does not bargain, does not cause—
It only speaks what I must feel.

172

I feel the ache behind my knees.
I feel the breath that takes its time.
I feel how fast the fire flees
From hands once swift, from limbs that climb.
And still I stand. And still I write.
Not to be loved or be adored—
But just to see, beneath the blight,
If something holy was ignored.

173

Ignored, not lost—perhaps it waits.
Not buried deep, but close, within.
Behind the locks and golden gates
I built to keep out love and sin.
Perhaps I feared not pain but peace—
What comes when hunger's finally still.
Perhaps I feared the great release
From needing more than one small will.

174

I ate the fruit of every tree
That whispered pleasure in my ear.
I drank the wine that made me flee
From every shadow shaped by fear.
But now the orchard's dead and dry,
And all that's left is root and stone.
Yet even here, beneath this sky,
A seed may crack when left alone.

175

Alone: the word that used to burn,
That sent me racing to a crowd.
But now I find in its return
A silence not so harsh or loud.
It does not mock, it does not steal.
It simply sits and asks me why.
And in that question, slow and real,
I almost glimpse a clearer sky.

176

I've seen the cities, touched the flesh.
I've woken rich and gone to waste.
I’ve felt the thrill of garments fresh,
Then thrown them out in breathless haste.
The world is not enough to fill
The hole that will not name its source.

For only love—not lust or will—

Can turn a raging stream to course.

177

My father died with quiet lips.

He left no speech, no final plea.

Just folded hands and slowing hips—

As if he knew what I can't see.

And now I wonder, was he wise

To never chase what made me bleed?

To live beneath more humble skies

And never feed this endless need?

178

I've read the scrolls, I've mocked the saints.

I've torn down altars, built my own.

I've painted pleasure in bright paints

While casting mercy out in stone.

But now I walk where churches lean,
Where graveyards whisper under trees,
And think perhaps I might have seen
The truth behind those bended knees.

179

Not fear—but awe. Not chains—but light.
Not guilt—but clarity and grace.
Not burning shame, but second sight
To see the soul in every face.
And if I can't reclaim the path,
Then let me clear it for the young—
To warn them how the golden wrath
Devours hearts before they're sung.

180

Let one small boy, with open eyes,
Not walk the steps that I have trod.

Let one young girl, where ruin lies,
Find roots instead, and seek her God.
Let someone read this crumbling page
And stop before the feast begins.
Let age become a kind old sage—
Not just the sum of hollow sins.

181

Let mercy meet me where I fall,
Not with a trumpet or a crown,
But in the dust beyond the wall—
Where all pretense is broken down.
Let me be poor, but wide awake.
Let me be small, but still aware.
For if I have a soul to break,
Then maybe I've a soul to spare.

182

I will not beg to change the past,
Nor ask the wind to blow it clean.
I'll let its lessons hold me fast,
Each crack and cut and cruel machine.
I'll wear the scars as honest things—
Not trophies, not excuses made—
But silent bells, with iron rings,
To mark the debts I never paid.

183

I once believed I'd leave a mark—
A statue, name, or gilded tale.
But now I see the truest spark
Is not what shines, but what prevails.
A kindness done, a lie withheld,
A meal made warm for one in need—
These shape the soul that can't be felled,
Though time may level every deed.

184

I laughed at men who bent to pray—
So sure the sky was void and still.
But I had never knelt that way,
Or dared to let go of my will.
I feared the silence might reply.
I feared the light would burn too true.
But now I kneel, and ask not why—
But only, “Is there more than you?”

185

Somewhere, a child becomes a thief.
Somewhere, a dealer makes a trade.
Somewhere, a mother weeps in grief
For promises her boy betrayed.
I’ve walked in each of these before—
The thief, the seller, and the son.

And now I stand before the door
Where all of these might be undone.

186

Undone—not by the force of shame,
Nor threats of hell or final breath.
But by the whisper in my name
That turns the edge of life from death.
Not justice sharp, but justice kind,
That seeks the truth beneath the bruise.
That sees what sickness did to mind,
And gives the damned a chance to choose.

187

I do not dream of paradise.
I do not crave a seat on high.
I only want to break the ice
That formed around a younger cry.

The boy I was still haunts the dust—

Still clutches hope in trembling fists.

And if I reach him, if I must,

Let him not vanish as he twists.

188

He twisted once from every hand

That offered love without a price.

He could not bear to understand

That not all mercy comes with vice.

I'd tell him now, if I still could,

That not all fire was born to burn.

That some will sit with him for good,

And ask for nothing in return.

189

But if no child remains in me,

No soft unguarded place to weep,

Then let me guard that memory

For those who still have wounds too deep.

Let me grow gentler with my rage.

Let me grow simpler in my creed.

Let me become the quiet sage

Who knows the shape of every need.

190

For I have known the hungers vast—

The ones no banquet ever fed.

And I have known the ghosts that last

Long after every joy has fled.

But if I walk with open eyes,

And if I speak without disguise,

Then even in my slow demise

I leave behind no further lies.

191

The bottle's voice is softer now.
The needle speaks a fading song.
The urge to dazzle, cheat, and plow
Through love is not as loud or strong.
But I remember every face
I hurt to feed the fire within.
I do not ask to be replaced—
I only ask to not begin.

192

Begin again? No, not from start.
I do not seek a second youth.
But maybe from a wiser heart
I might declare a harder truth.
That life is not a chase to win,
Nor stage for endless self-display—
But sacred ground we carry in,
And shape by what we give away.

193

I gave too little, asked too much.

I praised myself, ignored the cost.

And now, with age's subtle touch,

I feel the depth of what was lost.

But even loss is not the end—

If loss becomes the path to grace.

If I can weep, and kneel, and mend—

Then I am not in Hell’s embrace.

194

The fever's gone. The noise is dim.

The screens have ceased to flicker lies.

The altar built to self and whim

Has crumbled in the open skies.

And all that's left is breath and ground,

And one last dawn I did not earn.

If this is all I leave or found,

Then let it be my soul's return.

195

Return—not to a temple's gate,

Not to the chants or gilded verse.

But to the still, the small, the late—

The mercy offered to the worst.

For I was worst in many ways—

The liar, tyrant, fool, and thief.

But now I count my slow, gray days

As sacred spaces for belief.

196

Belief not in what I once claimed—

But in the spark that I betrayed.

Belief that love is never shamed

By how the broken life was played.

Belief that God, if He is real,

Can find me even now, unmasked.

And if He speaks, He speaks to heal—

Not judge, not burn, but gently ask.

197

Ask what I made, and what I spent.

Ask what I feared, and what I fled.

Ask what I meant in each lament,

And where I turned when wisdom bled.

Ask me not as warden, cold—

But like a father come too late.

A presence neither young nor old,

But simply full, and simply wait.

198

And if no answer comes at last,

No angel's voice or glowing sign,

Then let me sit within the past

And still make peace with what was mine.

For even silence, when it stays,

May carry weight, may be a vow.

And though I strayed in countless ways,

I am, at least, still here. Still now.

199

And if the page is soon to close—

If breath withdraws and light recedes—

Then let me lie in plain repose,

No boast, no claims, no frantic needs.

Let only one small truth remain:

That I once feasted, then I saw.

That though I dwelled in smoke and stain,

I longed, at last, for holy law.

200

The feast is done. The room is bare.

The mirror shows a quiet man.

No longer seeking, loud, or rare—

Just one who tried, as best he can,

To name the waste, to face the end,

To hold regret and still forgive.

To bless the wound, to love, to bend—

And let a broken soul still live.

Would you like a PDF or clean document of the full 200-stanza poem for download or printing?

A Sport of Ow

The whistle blows, the crowd roars loud,

As warriors take the field so proud.

But in this game of sweat and strain,

Each triumph comes with threads of pain.

The runner's lungs burn fierce and bright,

Legs screaming through the endless night.

The swimmer's shoulders cry for rest,

While chasing times to beat their best.

The boxer's knuckles split and bleed,

The cyclist's muscles plead and plead.

The gymnast's joints bend past their will,

Yet still they climb each painful hill.

For in this sport we've come to know,

Where victories and bruises grow,

The sweetest wins are often found

Where "ow" and "wow" share common ground.

So here's to those who choose to play

This game where hurting leads the way—

Who understand that strength is born

From dancing with each painful thorn.

The sport of ow, both harsh and true,

Reveals what mortals' hearts can do:

Transform each ache into a crown,

And never let the pain break down.

Gemini AppsPrompted long gritty urban poem about all the things people do to sicken themselves with consumption and waste in lives that have little meaning but for greed, anger, gluttony, narcissism, hedonism, and immediate gratification, as well as selfishness and Godlessness drug use, drunkenness, carelessness, self abuse. about all the things people do to sicken themselves with consumption and waste in lives that have little meaning but for greed, comfort, immediate gratification, and hedonism, gambling, addiction, distractionDetailseventJuly 11 at 9:02 PMappsGemini Appschat

Concrete Rot: A Modern City Symphony of Self-Abuse

The city breathes, a smog-choked, coughing sigh, Beneath a digital, unblinking eye. Glass towers pierce the bruised and purple dawn, Where restless ghosts of meaning are withdrawn. A thousand screens ignite, a silent hum, No gentle wakening, just the urgent drum Of instant news, of phantom, viral dread, Injecting fear, inside each waking head. The phone a tether, pulling mind from soul, Surrendering to algorithms' control.

The first harsh cough, of smoke-filled, stale air drawn, Another hit before the day is born. The nervous twitch, a craving deep and cold, For chemical solace, stories left untold. The needle's gleam, a momentary peace, From nagging truths, from pain that won't release. A fleeting high, a journey far from real, While life outside, its bitter wounds reveal. The skin too pale, the eyes a vacant stare, Lost to the shadows, caught in self-made snare.

The coffee's bitter, scalding, in the hand, Fuel for the frantic, across the land. A disposable cup, a symbol of our haste, Another fragment, carelessly laid to waste. The sugary pastry, sweet and sickly rich, A glutton's start, a self-destructive hitch. No nourishment, no thoughtful, mindful bite, Just hurried stuffing, in the morning's light. The stomach churns, a dull and constant ache, For pleasures taken, for convenience's sake.

The steel-cased grind, of metal on the rail, Commuters numb, a silent, tired wail. Faces reflected, grim and set and cold, A thousand stories, waiting to unfold. But eyes are fixed on screens, a hypnotic gleam, Lost in a world, a self-created dream. No human contact, no acknowledged nod, Just individual silos, blessed by God Of disconnection, chosen and maintained, A vast indifference, perfectly ingrained.

The office towers, canyons stark and tall, Where ambition's echo answers every call. The frantic emails, flashing, sharp, and fast, A thousand targets, meant to always last. The greed for power, clawing, bitter, raw, Obeying profits, bending every law. The backroom deals, the whispers of deceit, As ethics wither, in the sterile heat Of endless striving, for the corner suite, A hollow victory, tasting bittersweet.

Then break descends, a blur of hurried feet, Towards the solace of the city street. The vape pen rises, with its scented cloud, Another habit, shouted to the crowd. The nicotine's grip, a subtle, binding chain, A momentary comfort, easing inner pain. The fleeting urge, addressed with practiced hand, While self-abuse spreads subtly through the land. A slow decay, ignored in daily quest, For easy pleasure, putting lungs to test.

Lunch is a blur, a phone in one moist palm, Devouring content, safe from inner storm. The fast-food wrapper, slick with greasy shame, Another indulgence, played a silent game. The portions swollen, piled beyond all need, A testament to gluttony's dark creed. The body protests, sluggish, full, and ill, But urges whisper, "Eat a little still." Then tossed aside, the remnants quickly found, A mountain growing, on the city ground.

The afternoon, a drone of discontent, On tasks robotic, where our souls are lent To corporate whims, to profit's cruel command, While vibrant passions wither in the sand. The inner voice, a whisper, soft and low, Of gifts unshared, of seeds we failed to sow. But comfort's chains, and greed's seductive hold, Keep spirits captive, as the days grow old. The years evaporate, a vapor, thin and faint, A painted smile, concealing true complaint.

The children's rooms, a riot of excess, Toys bought and broken, causing endless stress. The screens ablaze, with games and vivid hues, Consuming childhood, shaping future views. No simple play, no quiet, patient craft, Just instant craving, on a digital raft. The latest gadget, fleeting, quick, and new, To soothe the temper, to make dreams come true. But true contentment, born of time and touch, Is lost to wanting, having far too much.

The dinner hour, a solitary show, Before the screen, where endless dramas flow. No shared communion, no engaging plea, Just silent chewing, for all eyes to see The virtual feast, while real-life ties grow weak, A lonely comfort, no true words to speak. The takeout boxes, piling in hasty waste, The flavors fleeting, lost to urgent haste. The art of cooking, bonding, slow and deep, Replaced by numbness, while the senses sleep.

The evening rolls, a binge of mindless bliss, A thousand plots, a stolen, fleeting kiss Of false escape, from worries, real and vast, Till tired eyes surrender, at the last. The body stiffens, mind begins to blur, The soul's deep yearning, left without a stir. A digital drug, a potent, numbing hand, Across the conscience, spreading through the land. The precious hours, vanished in the night, For fleeting pleasure, dimming inner light.

The cocktail flows, a liquid, fiery grace, To drown the doubts, to numb the weary face. Another round, another fleeting high, As drunkenness takes hold, and limits die. The slurred confessions, bitterness released, A fragile ego, from restraint decreased. The morning after, shame and dull regret, A toxic ritual, deeply, firmly set. A hedonist's pursuit, a frantic quest, For fleeting joy, putting soul to test.

The bathroom gleams, with bottles, tubes, and jars, A daily ritual, fighting time's harsh scars. The endless masks, for beauty's cruel facade, A painted promise, on a spirit sad. Narcissus gazing, at a fading form, While inner beauty weathers every storm. The fear of aging, etched upon the mind, A futile battle, endlessly defined

By perfect skin, and hair of vibrant hue, Ignoring wisdom, choosing what is new.

The luxury car, a gleaming, metal shrine, A symbol grand, of power, truly mine. The envy sought, the status bravely won, As humble virtues wither in the sun. The silent judging, from a lofty seat, Of those less fortunate, beneath our feet. The roar of engines, drowning out the plea, For shared compassion, for humanity. A gilded cage, where selfishness resides, And genuine connection softly hides.

The sprawling house, a fortress built of greed, More rooms than needed, planted like a seed. To fill the emptiness, with bought delight, A silent yearning, in the lonely night. Each vacant chamber, echoing the lack, Of purpose deeper, setting life off track. The lavish spaces, rarely truly lived, A testament to what we truly grieved— The loss of meaning, in the grand design, As comfort's prison, truly feels divine.

The latest gadget, flashing, sleek, and bright, A whispered promise, of enduring light. The older model, scorned and cast aside, Though still quite able, pushed out by the tide Of planned obsolescence, cleverly designed, To keep us wanting, leaving peace behind. The e-waste mountains, toxic, vast, and grim, A bitter legacy, up to the brim. A digital addiction, hard to truly break, For instant gratification, for its sake.

The fleeting trips, by air, a silver streak, For weekend escapes, where pleasures we can seek. The distant lands, reduced to hurried tours, Through crowded markets, passing famous doors. A selfie taken, proof that we were there, But deeper culture lost, beyond compare. The carbon footprint, soaring to the sky, As fleeting moments quickly pass us by. The world reduced to photos, quick and fast, A shallow journey, never meant to last.

The health trends bloom, a promise soft and sweet, Of perfect bodies, from our head to feet. The quick-fix diets, supplements untold, A dizzying array, bravely bought and sold. We chase perfection, starve and then we feast, A punishing cycle, never quite at peace. Ignoring wisdom, from within our core, For quick solutions, wanting ever more. The body aches, from comforts overdone, The race for beauty, never truly won.

The credit card, a powerful, plastic key, To open doors to instant ecstasy. The future mortgaged, for a present thrill, A rising balance, waiting to fulfill The empty promise, of a life well-spent, On fleeting pleasures, ultimately lent. The gnawing debt, a silent, heavy chain, That binds the spirit, bringing subtle pain. The freedom lost, to truly make a choice, Drowned out by interest's demanding voice.

The casual gifts, a token, quickly bought, For social circles, lessons never taught. Of thoughtful gestures, true connection's art, Replaced by objects, playing out a part. The packaging torn, the item put away, A fleeting moment, lasting but a day. The act of giving, stripped of deeper grace, Just one more thing, in this consumer race. A hollow gesture, easy, quick, and plain, Reflecting back, the emptiness again.

The entertainment yearns, a constant drone, From background noise, when we are quite alone. The music streams, a thousand songs on tap, But quiet contemplation starts to nap. No single note, allowed to truly breathe, Just endless options, a chaotic wreath. The brain adjusts, to constant, swift demand, Losing the power to truly understand The subtle beauty, of a single sound, In constant seeking, nothing deep is found.

The news addiction, fed by fright and dread, Of every crisis, fear within the head. We click and share, the outrage and the blame, A self-inflicted, agonizing game. Consuming sorrow, fueling inner strife, Distracting from the beauty of true life. The world's deep wounds, absorbed without a pause, Then quickly forgotten, by commercial laws. A shallow sympathy, a fleeting tear, Lost in the deluge, drowning out what's clear.

The storage units, filled to bursting height, With objects banished from our daily sight. The things we bought, then quickly cast aside, A monument to want, where treasures hide. Another rent payment, for a distant room, To hold the remnants of consumption's gloom. A silent burden, rarely understood, The cost of ownership, rarely for our good. We fill our lives, with what we don't quite need, And plant the seeds of future, silent greed.

The plastic bottles, in a daily stream, The single-use items, a forgotten dream Of less consumption, simple, plain, and pure, Replaced by convenience, for which we endure The growing mountains, in a landfill vast, A bitter echo, meant to truly last. The ocean's breath, now choked by human hand, A poisoned legacy, across the land. Each wrapper tossed, a tiny, careless plea, For a world burdened, endlessly to be.

The quest for status, a relentless chase, To show the world our self-appointed place. The brands we wear, the labels we display, To signify our worth, come what may. A fragile image, built on bought esteem, A hollow promise, like a waking dream. The true connection, based on heart and soul, Is lost to pretense, losing all control. We seek approval, in the envious eye, And let our deeper, truer essence die.

The hurried meals, consumed before a screen, No mindful eating, not a quiet scene. The flavors lost, the texture barely felt, As nourishment becomes a burden dealt. The body's signals, ignored and left unheard, As quick indulgence becomes the only word. Then bloated bellies, sluggish minds arise, The price of speed, reflected in our eyes. A vital act, reduced to mere intake, For instant pleasure, for convenience's sake.

The constant noise, a symphony of need, From advertisements, sowing restless seed. "Buy this," "Buy that," the whispers never cease, Disrupting quiet, stealing inner peace. They shape our longings, mould our very thought, Till what we crave, is precisely what they've taught. The independent spirit starts to wane, Enslaved by urges, caught in golden rain Of shiny objects, promises so grand, A marionette, within their clever hand.

The fear of missing out, a potent sting, Drives hurried choices, making spirits sing A fleeting anthem, of what must be had, Before the moment passes, leaving sad And lonely feelings, of being left behind, By fleeting trends, by pleasures hard to find. We join the rush, though knowing it's a lie, A desperate yearning, beneath a watchful sky. A social pressure, hard to stand against, For genuine connection, deeply fenced.

The quest for comfort, an insatiable quest, No minor hardship, putting us to test. The perfect temperature, the softest bed, The easiest path, where nothing's left unsaid About our wants, our needs, our smallest whim, Until the spirit's strength grows weak and dim. The joy of challenge, overcome with grace, Is lost to ease, in comfort's soft embrace. A pampered life, a body growing frail, Missing the lessons, carried on the gale.

The restless sleep, disturbed by digital glow, The mind still racing, putting on a show Of endless thoughts, of purchases unseen, Of futures planned, on a digital screen. The body aches, from habits gone astray, The price of comfort, at the close of day. No true repose, no deep, replenishing rest, Just weary slumber, putting us to test. The vital energy, consumed and spent, On meaningless pursuits, divinely sent.

The spirit yearns, for meaning, deep and wide, Beyond the surface, where illusions hide. A quiet voice, a whisper from within, Against the clamor, where the battles begin. To make, to build, to share, to truly be, To find contentment, in simplicity. To give oneself, in service, true and bold, A richer story, waiting to unfold.

But drowned by urges, trivial and fast, The soul's deep wisdom, fades into the past.

The friendships wane, for lack of time and care, Exchanged for chasing pleasures in the air. The long conversations, heart to heart and true, Are lost to scrolling, for a hurried view. The shared experience, profound and deeply felt, Is sacrificed, for comforts quickly dealt. The human touch, so vital, kind, and warm, Is swept away by consumption's passing storm. A lonely journey, filled with bought delight, But missing solace, in the fading light.

The earth cries out, a silent, painful plea, From ravaged landscapes, for what used to be. The poisoned rivers, choking, dark, and wide, Reflect the carelessness, we cannot hide. The air grows heavy, with a silent dread, As future generations, lie unbred. Our wanton hunger, leaves a barren land, A bitter legacy, by human hand. The natural world, a treasure left behind, For fleeting comforts, of a grasping mind.

The empty purpose, driven by the 'me', A selfish journey, for the eye to see. No higher calling, no collective aim, Just ego's whisper, playing out its game. The larger questions, of our common plight, Are pushed aside, for personal delight. A life consumed, by what the self desires, Ignoring fires, setting further fires. The fleeting pleasure, paramount and grand, While greater purpose slips from out our hand.

The quiet mornings, never truly felt, The sunbeams wasted, as the spirit knelt To fleeting cravings, instant, sharp, and brief, A constant seeking, finding no relief. The simple beauty, of a bird in flight, Is lost to distractions, in the harsh glare of light. The gentle rhythm, of a slower pace, Erased by rushing, in this endless race. A world of wonder, vibrant, rich, and deep, While senses slumber, and our spirits sleep.

The constant wanting, never satisfied, A restless craving, always deep inside. The thrill of purchase, quickly fades away, Leaving a void, at the close of every day. The hedonic treadmill, spinning ever fast, No lasting joy, no happiness to cast A steady anchor, in a stormy sea, Just constant striving, for what cannot be. A promise broken, whispered soft and low, That more possessions, make the spirit glow.

The health declines, a subtle, creeping ache, For wellness sacrificed, for pleasure's sake. The hurried meals, the lack of restful sleep, The body's protest, secrets it will keep Until the breaking, when the system strains, And vital energy, slowly starts to wane. The fleeting pleasures, leave a bitter tang, As wellness falters, like a broken pang. The body's temple, ravaged and laid low, By habits careless, and a steady flow.

The spirit wanes, a dull and weary sigh, As deep connections are allowed to die. No soulful moments, shared with open heart, Just superficial roles, to play a part. The bonds of family, stretched and growing thin, As individual comforts pull us in. The community's call, a distant, muffled sound, As self-absorption, circles all around. A lonely island, built on bought delight, But missing solace, in the fading light.

The shallow laughs, that echo in the air, Reflecting emptiness, beyond compare. No genuine joy, from purpose deeply found, Just fleeting moments, on superficial ground. The forced enthusiasm, bright and bold, A weary story, constantly retold. The search for happiness, in things outside, A treasure hidden, where true feelings hide. A life of pretense, lived for outward show, While inner emptiness continues to grow.

The future dims, a hazy, uncertain view, As present comforts, blind us to what's true. The warnings echo, from a distant shore, But hedonism always asks for more. The planet groans, the symptoms multiply, Yet we consume, beneath a silent sky. The selfish bubble, vibrant, strong, and bright, Until it bursts, in overwhelming light. A stark awakening, cruel and cold and grim, When future consequences reach the rim.

The inner voice, a whisper, soft and low, Of chances missed, of seeds we failed to sow. The talents buried, aspirations drowned, By endless wanting, on consumer ground. The purpose lost, the meaning left behind, In constant seeking, for a restless mind. A life defined by what we took and bought, A hollow legacy, by conscience taught. The quiet knowing, in the lonely night, Of wasted moments, and extinguished light.

The bitter harvest, gathered at the end, No rich fulfillment, no true, faithful friend To bear witness, to a life well-spent, Just fleeting pleasures, carelessly lent. The sum total: not joy, but lingering pain, A restless spirit, crying out in vain. For what is life, when all is stripped away, But what we gave, and lived for, every day? And if that giving, was to self alone, Then bitter emptiness becomes our throne.

The gambler's fever, hot within the veins, A desperate hope, through losses and through pains. The spinning wheel, the cards dealt, sharp and fast, A fantasy of fortune, never meant to last. The borrowed coins, the promises denied, As addiction's grip takes hold, deep inside. A quick sensation, thrilling, wild, and free, But leaves the spirit bound, eternally. The shattered trust, the broken, empty gaze, Lost to the numbers, in a money maze.

The distraction's hum, a constant, subtle pull, From every thought, making the spirit dull. The fleeting scroll, the endless, mindless game, A thousand small escapes, murmuring our name. No time for stillness, contemplation's art, Just fragmented moments, tearing us apart. The inner world, a landscape left untilled, While superficial desires are constantly fulfilled. A life unexamined, rushing past in haste, The precious moments, deliberately laid to waste.

The final tally, as the shadows fall, A wasted journey, answering greed's call. No lasting peace, no deep, abiding grace, Just emptiness reflected in our face. The spirit's hunger, never truly fed, By all the comforts, bought, and quickly shed. The bitter knowledge, heavy, cold, and vast, A life unlived, a future overcast. The broken promise, whispered in the air, Of what we might have been, beyond compare.

write a poem with the title "A Sport of Ow"EditA Sport of OwDocument #

A Sport of Ow

The whistle blows, the crowd roars loud,

As warriors take the field so proud.

But in this game of sweat and strain,

Each triumph comes with threads of pain.

The runner's lungs burn fierce and bright,

Legs screaming through the endleI've written a poem called "A Sport of Ow" that explores the relationship between pain and athletic achievement. The poem captures how physical discomfort is an inherent part of competitive sports, yet athletes embrace it as part of their journey toward excellence. It touches on various sports and the specific ways each one challenges the body, while celebrating the resilience of those who push through pain to achieve their goals.

The Great Unraveling: A Modern Reckoning

The digital dawn, a screen's anointing glow, Not gentle light, but data's endless flow. Before the sun has kissed the sleeping street, A thousand cravings, at our restless feet. We scroll and tap, in rituals profane, For instant fixes, easing phantom pain. The curated feed, a mirror to our greed, Planting the bitter, discontented seed. No quiet rising, no serene embrace, Just hurried hunger, etched upon the face.

The morning brew, a pod in plastic spun, Consumed and tossed, before the day's begun. A fleeting warmth, a jolt of borrowed zest, Then waste amassed, by comfort's endless quest. The hurried breakfast, processed, sweet, and bland, Each bite devoid of blessing from the land.

No conscious thanks, no gratitude profound, Just eager haste, on hollow, wasteful ground. The easy choice, the effortless delight, Obscuring consequence, by morning light.

The commute begins, a single, captive soul, In steel and glass, surrendering control. The roaring engine, burning ancient might, Propelling forward, blind to future's plight. Miles logged in isolation, comfort's plea, While public pathways beckon, strong and free. The planet groans, a whisper in the air, Of warming currents, burdened by our care— Or lack thereof, as fumes ascend and spread, For selfish journeys, where the spirit's dead.

The closet bursts, a monument to want, Each folded garment, whispers, "More, not scant!" A tyrant trend, dictating what to wear, While sweatshop echoes linger in the air. Fast fashion's lure, a cheap and fleeting game, A thousand outfits, whispering our name. Worn once, perhaps, then banished from the light, To landfill mountains, stretching through the night. The human cost, the planet's silent plea, Lost in the fabric of our vanity.

The kitchen gleams, with gadgets seldom used, Each specialized, a promise long diffused. The air fryer, blender, spiralizer's might, Bought on a whim, then banished from the light. Processed foods, in bright, deceptive guise, Adorn the shelves, where real nutrition dies. A thousand flavors, engineered to please, But leave the body weak, and ill at ease. The fresh produce wilts, a quiet, sad rebuke, As easy meals become the daily fluke.

The office gleams, a gilded, sterile cage, Where human purpose turns another page. The frantic emails, flashing, cold, and fast, A thousand targets, meant to always last. We chase the bonus, seek the corner room, A ceaseless striving, swallowed by the gloom Of empty victories, hollow, bought, and sold, A life unwritten, a story left untold. The climb for power, anger's subtle sting, As ruthless ambition makes the spirit sing.

The break descends, a hurried, screen-lit space, Another click, another empty chase. The online deals, the virtual delight, Consuming minutes, in the fading light. A brief escape, from tasks we can't endure, To mindless Browse, promises impure. The instant thrill, a dopamine quick fix, As real connection withers, sadly kicks Against the glass, unheard, unheeded, stark, Leaving the spirit empty, in the dark.

Then lunch arrives, in plastic, bright and bold, A meal devoured, a story left untold. Of nourishment, of mindful, gentle chew, Replaced by haste, and thoughts of what's anew. The fork attacks, the portions piled too high, A glutton's feast beneath an uncaring sky. The body stuffed, protesting, feeling ill, Yet urges whisper, "Eat a little still." The waste discarded, food that could have fed, A silent sin, upon the table spread.

The afternoon, a drone of discontent, On tasks robotic, where our souls are lent To corporate whims, to profit's endless cry, While dreams of passion quietly pass by. The inner voice, a whisper, soft and low, Of gifts unshared, of seeds we failed to sow. But greed's sharp teeth, and comfort's easy hold, Keep spirits captive, as the days grow old. The years evaporate, a vapor, thin and faint, A painted smile, concealing true complaint.

The children's rooms, a riot of excess, Toys bought and broken, causing endless stress. The screens ablaze, with games and vivid hues, Consuming childhood, shaping future views. No simple play, no quiet, patient craft, Just instant craving, on a digital raft. The latest gadget, fleeting, quick, and new, To soothe the temper, to make dreams come true. But true contentment, born of time and touch, Is lost to wanting, having far too much.

The dinner hour, a solitary show, Before the screen, where endless dramas flow. No shared communion, no engaging plea, Just silent chewing, for all eyes to see The virtual feast, while real-life ties grow weak, A lonely comfort, no true words to speak. The takeout boxes, piling in hasty waste, The flavors fleeting, lost to urgent haste. The art of cooking, bonding, slow and deep, Replaced by numbness, while the senses sleep.

The evening rolls, a binge of mindless bliss, A thousand plots, a stolen, fleeting kiss Of false escape, from worries, real and vast, Till tired eyes surrender, at the last. The body stiffens, mind begins to blur, The soul's deep yearning, left without a stir. A digital drug, a potent, numbing hand, Across the conscience, spreading through the land. The precious hours, vanished in the night, For fleeting pleasure, dimming inner light.

The cocktail flows, a liquid, fiery grace, To drown the doubts, to numb the weary face. Another round, another fleeting high, As self-control begins to gently die. The slurred confessions, bitterness released, A fragile ego, from restraint decreased. The morning after, shame and dull regret, A toxic ritual, deeply, firmly set. A hedonist's pursuit, a frantic quest, For fleeting joy, putting soul to test.

The bathroom shelves, a cluttered, vain display, Of lotions, potions, washing woes away. The endless masks, for beauty's cruel facade, A painted promise, on a spirit sad. Narcissus gazing, at a fading form, While inner beauty weathers every storm. The fear of aging, etched upon the mind, A futile battle, endlessly defined By perfect skin, and hair of vibrant hue, Ignoring wisdom, choosing what is new.

The luxury car, a gleaming, metal shrine, A symbol grand, of power, truly mine. The envy sought, the status bravely won, As humble virtues wither in the sun. The silent judging, from a lofty seat, Of those less fortunate, beneath our feet. The roar of engines, drowning out the plea, For shared compassion, for humanity. A gilded cage, where selfishness resides, And genuine connection softly hides.

The sprawling house, a fortress built of greed, More rooms than needed, planted like a seed. To fill the emptiness, with bought delight, A silent yearning, in the lonely night. Each vacant chamber, echoing the lack, Of purpose deeper, setting life off track. The lavish spaces, rarely truly lived, A testament to what we truly grieved— The loss of meaning, in the grand design, As comfort's prison, truly feels divine.

The latest gadget, flashing, sleek, and bright, A whispered promise, of enduring light. The older model, scorned and cast aside, Though still quite able, pushed out by the tide Of planned obsolescence, cleverly designed, To keep us wanting, leaving peace behind. The e-waste mountains, toxic, vast, and grim, A bitter legacy, up to the brim. A digital addiction, hard to truly break, For instant gratification, for its sake.

The fleeting trips, by air, a silver streak, For weekend escapes, where pleasures we can seek. The distant lands, reduced to hurried tours, Through crowded markets, passing famous doors. A selfie taken, proof that we were there, But deeper culture lost, beyond compare. The carbon footprint, soaring to the sky, As fleeting moments quickly pass us by. The world reduced to photos, quick and fast, A shallow journey, never meant to last.

The health trends bloom, a promise soft and sweet, Of perfect bodies, from our head to feet. The quick-fix diets, supplements untold, A dizzying array, bravely bought and sold. We chase perfection, starve and then we feast, A punishing cycle, never quite at peace. Ignoring wisdom, from within our core, For quick solutions, wanting ever more. The body aches, from comforts overdone, The race for beauty, never truly won.

The credit card, a powerful, plastic key, To open doors to instant ecstasy. The future mortgaged, for a present thrill, A rising balance, waiting to fulfill The empty promise, of a life well-spent, On fleeting pleasures, ultimately lent. The gnawing debt, a silent, heavy chain, That binds the spirit, bringing subtle pain. The freedom lost, to truly make a choice, Drowned out by interest's demanding voice.

The casual gifts, a token, quickly bought, For social circles, lessons never taught. Of thoughtful gestures, true connection's art, Replaced by objects, playing out a part. The packaging torn, the item put away, A fleeting moment, lasting but a day. The act of giving, stripped of deeper grace, Just one more thing, in this consumer race. A hollow gesture, easy, quick, and plain, Reflecting back, the emptiness again.

The entertainment yearns, a constant drone, From background noise, when we are quite alone. The music streams, a thousand songs on tap, But quiet contemplation starts to nap. No single note, allowed to truly breathe, Just endless options, a chaotic wreath. The brain adjusts, to constant, swift demand, Losing the power to truly understand The subtle beauty, of a single sound, In constant seeking, nothing deep is found.

The news addiction, fed by fright and dread, Of every crisis, fear within the head. We click and share, the outrage and the blame, A self-inflicted, agonizing game. Consuming sorrow, fueling inner strife, Distracting from the beauty of true life. The world's deep wounds, absorbed without a pause, Then quickly forgotten, by commercial laws. A shallow sympathy, a fleeting tear, Lost in the deluge, drowning out what's clear.

The storage units, filled to bursting height, With objects banished from our daily sight. The things we bought, then quickly cast aside, A monument to want, where treasures hide. Another rent payment, for a distant room, To hold the remnants of consumption's gloom. A silent burden, rarely understood, The cost of ownership, rarely for our good. We fill our lives, with what we don't quite need, And plant the seeds of future, silent greed.

The plastic bottles, in a daily stream, The single-use items, a forgotten dream Of less consumption, simple, plain, and pure, Replaced by convenience, for which we endure The growing mountains, in a landfill vast, A bitter echo, meant to truly last. The ocean's breath, now choked by human hand, A poisoned legacy, across the land. Each wrapper tossed, a tiny, careless plea, For a world burdened, endlessly to be.

The quest for status, a relentless chase, To show the world our self-appointed place. The brands we wear, the labels we display, To signify our worth, come what may. A fragile image, built on bought esteem, A hollow promise, like a waking dream. The true connection, based on heart and soul, Is lost to pretense, losing all control. We seek approval, in the envious eye, And let our deeper, truer essence die.

The hurried meals, consumed before a screen, No mindful eating, not a quiet scene. The flavors lost, the texture barely felt, As nourishment becomes a burden dealt. The body's signals, ignored and left unheard, As quick indulgence becomes the only word. Then bloated bellies, sluggish minds arise, The price of speed, reflected in our eyes. A vital act, reduced to mere intake, For instant pleasure, for convenience's sake.

The constant noise, a symphony of need, From advertisements, sowing restless seed. "Buy this," "Buy that," the whispers never cease, Disrupting quiet, stealing inner peace. They shape our longings, mould our very thought, Till what we crave, is precisely what they've taught. The independent spirit starts to wane, Enslaved by urges, caught in golden rain Of shiny objects, promises so grand, A marionette, within their clever hand.

The fear of missing out, a potent sting, Drives hurried choices, making spirits sing A fleeting anthem, of what must be had, Before the moment passes, leaving sad And lonely feelings, of being left behind, By fleeting trends, by pleasures hard to find. We join the rush, though knowing it's a lie, A desperate yearning, beneath a watchful sky. A social pressure, hard to stand against, For genuine connection, deeply fenced.

The quest for comfort, an insatiable quest, No minor hardship, putting us to test. The perfect temperature, the softest bed, The easiest path, where nothing's left unsaid About our wants, our needs, our smallest whim, Until the spirit's strength grows weak and dim. The joy of challenge, overcome with grace, Is lost to ease, in comfort's soft embrace. A pampered life, a body growing frail, Missing the lessons, carried on the gale.

The restless sleep, disturbed by digital glow, The mind still racing, putting on a show Of endless thoughts, of purchases unseen, Of futures planned, on a digital screen. The body aches, from habits gone astray, The price of comfort, at the close of day. No true repose, no deep, replenishing rest, Just weary slumber, putting us to test. The vital energy, consumed and spent, On meaningless pursuits, divinely sent.

The spirit yearns, for meaning, deep and wide, Beyond the surface, where illusions hide. A quiet voice, a whisper from within, Against the clamor, where the battles begin. To make, to build, to share, to truly be, To find contentment, in simplicity. To give oneself, in service, true and bold, A richer story, waiting to unfold. But drowned by urges, trivial and fast, The soul's deep wisdom, fades into the past.

The friendships wane, for lack of time and care, Exchanged for chasing pleasures in the air. The long conversations, heart to heart and true, Are lost to scrolling, for a hurried view. The shared experience, profound and deeply felt, Is sacrificed, for comforts quickly dealt. The human touch, so vital, kind, and warm, Is swept away by consumption's passing storm. A lonely journey, filled with bought delight, But missing solace, in the fading light.

The earth cries out, a silent, painful plea, From ravaged landscapes, for what used to be. The poisoned rivers, choking, dark, and wide, Reflect the carelessness, we cannot hide. The air grows heavy, with a silent dread, As future generations, lie unbred. Our wanton hunger, leaves a barren land, A bitter legacy, by human hand. The natural world, a treasure left behind, For fleeting comforts, of a grasping mind.

The empty purpose, driven by the 'me', A selfish journey, for the eye to see. No higher calling, no collective aim, Just ego's whisper, playing out its game. The larger questions, of our common plight, Are pushed aside, for personal delight. A life consumed, by what the self desires, Ignoring fires, setting further fires. The fleeting pleasure, paramount and grand, While greater purpose slips from out our hand.

The quiet mornings, never truly felt, The sunbeams wasted, as the spirit knelt To fleeting cravings, instant, sharp, and brief, A constant seeking, finding no relief. The simple beauty, of a bird in flight, Is lost to distractions, in the harsh glare of light. The gentle rhythm, of a slower pace, Erased by rushing, in this endless race. A world of wonder, vibrant, rich, and deep, While senses slumber, and our spirits sleep.

The constant wanting, never satisfied, A restless craving, always deep inside. The thrill of purchase, quickly fades away, Leaving a void, at the close of every day. The hedonic treadmill, spinning ever fast, No lasting joy, no happiness to cast A steady anchor, in a stormy sea, Just constant striving, for what cannot be. A promise broken, whispered soft and low, That more possessions, make the spirit glow.

The health declines, a subtle, creeping ache, For wellness sacrificed, for pleasure's sake. The hurried meals, the lack of restful sleep, The body's protest, secrets it will keep Until the breaking, when the system strains, And vital energy, slowly starts to wane. The fleeting pleasures, leave a bitter tang, As wellness falters, like a broken pang. The body's temple, ravaged and laid low, By habits careless, and a steady flow.

The spirit wanes, a dull and weary sigh, As deep connections are allowed to die. No soulful moments, shared with open heart, Just superficial roles, to play a part. The bonds of family, stretched and growing thin, As individual comforts pull us in. The community's call, a distant, muffled sound, As self-absorption, circles all around. A lonely island, built on bought delight, But missing solace, in the fading light.

The shallow laughs, that echo in the air, Reflecting emptiness, beyond compare. No genuine joy, from purpose deeply found, Just fleeting moments, on superficial ground. The forced enthusiasm, bright and bold, A weary story, constantly retold. The search for happiness, in things outside, A treasure hidden, where true feelings hide. A life of pretense, lived for outward show, While inner emptiness continues to grow.

The future dims, a hazy, uncertain view, As present comforts, blind us to what's true. The warnings echo, from a distant shore, But hedonism always asks for more. The planet groans, the symptoms multiply, Yet we consume, beneath a silent sky. The selfish bubble, vibrant, strong, and bright, Until it bursts, in overwhelming light. A stark awakening, cruel and cold and grim, When future consequences reach the rim.

The inner voice, a whisper, soft and low, Of chances missed, of seeds we failed to sow. The talents buried, aspirations drowned, By endless wanting, on consumer ground. The purpose lost, the meaning left behind, In constant seeking, for a restless mind. A life defined by what we took and bought, A hollow legacy, by conscience taught. The quiet knowing, in the lonely night, Of wasted moments, and extinguished light.

The bitter harvest, gathered at the end, No rich fulfillment, no true, faithful friend To bear witness, to a life well-spent, Just fleeting pleasures, carelessly lent. The sum total: not joy, but lingering pain, A restless spirit, crying out in vain. For what is life, when all is stripped away, But what we gave, and lived for, every day? And if that giving, was to self alone, Then bitter emptiness becomes our throne.

Gemini Apps

Prompted long poem of at least 50 stanzas about all the things people do to sicken themselves with consumption and waste in lives that have little meaning but for greed, anger, gluttony, narcissism, hedonism, and immediate gratification, as well as selfishness and Godlessness

apps

Gemini Apps

chat

The Feast of Emptiness

The modern table, laden, groans and sighs, Not with true plenty, but with gilded lies. We gather 'round, a restless, hungry throng, And gorge ourselves where we can't truly belong. A world of plenty, spun for every whim, A constant feast, where meaning grows so dim.

The Daily Ritual: A Toxic Tide

The sun ascends, and with it, hunger stirs, Not for true sustenance, but the fleeting whirs Of coffee pods, in plastic-laden guise, A bitter brew, beneath indifferent skies. Then packaged bread, in wrappers bright and bold, Each hurried bite, a story left untold Of fertile soil, of labor, patience, care, Replaced by speed, and waste beyond compare. The car, alone, a metal, hungry beast, Consuming fuel for every meager feast Of urban sprawl, of errands swiftly run, While bicycles rust, beneath a weary sun. The workday hums, a drone of discontent, For wages earned, on comforts quickly spent. More clothes online, a click, a fleeting thought, Another garment, instantly forgot, As fashion spins, in cycles fast and crude, A restless spirit, constantly pursued By endless newness, never quite attained, A satisfaction, endlessly constrained. The lunch arrives, in styrofoam and plastic sheen, A meal consumed, a momentary scene, Then tossed aside, a monument to haste, Another burden, on a planet laid to waste. The evening calls, with screens that brightly gleam, A digital feast, a captivating dream. We scroll and swipe, for dopamine's quick hit, Our minds devoured, by the bytes we sit And passively absorb, till eyes grow numb, And real connection's echo fails to come. Then takeout boxes, piling in the bin, A final act, before the night draws in. The air-con blasts, the heating hums and roars, Ignoring weather, closing all the doors On natural rhythms, balanced, calm, and slow, Demanding comfort, letting nature go.

The Sickness Spreads: Body, Mind, and Earth

Our bodies ache, from comforts over-fed, The processed sugar, heavy, like a dread That settles deep, in organs strained and sore, As vibrant health knocks feebly at the door. The quick fix sought, in pills and fleeting fads, Ignoring whispers from the wisdom ads Of moderation, movement, quiet rest, Exchanged for ease, a self-inflicted test. Our minds grow thin, distracted, over-stimulated, By constant noise, our true selves now belated. The joy of crafting, making, building, learning, Replaced by buying, borrowing, always yearning. We fill the void with things, a desperate plea, For meaning lost, in sheer futility. The spirit starved, for purpose, deep and profound, While endless goods heap silent, all around. The earth, meanwhile, a fevered, trembling form, Endures the brunt of every consumption storm. The forests fall, for paper, fleeting wood, The oceans choke, where plastic once withstood The tests of time, now broken, scattered wide, A bitter legacy, we cannot hide. The air grows thick with fumes of restless flight, Our burning wants, eclipsing stars at night. Resource depleted, poisoned, scarred, and torn, A legacy of shame, for those unborn.

The Empty Legacy: A Life Unlived

And in the end, when shadows start to creep, And scattered remnants, silent vigil keep, What will remain of all the things we sought? A fleeting pleasure, dearly, dearly bought. The sprawling house, filled with forgotten things, The silenced laughter, that the memory brings Of moments squandered, on a shallow quest, For comfort's lie, for putting self to test Against the mirror of another's gain, A ceaseless struggle, fueled by self-made pain. The meaning lost, in chasing shiny toys, The soul's deep yearning, drowned in trivial joys. A life consumed, by consuming's empty call, Leaving a legacy of nothing, after all.

The Unlived Lives: A Taxonomy of Regret

The Gilded Cage A life spent polishing a shining chain, For others' praise, or to avoid their pain. The doctor, stifled, yearned to paint and draw, The artist, boxed within the merchant's law. They walked the path their parents carved and laid, And in that chosen comfort, slowly swayed From their own truth, till yearning turned to rust, A silent scream, unheard, within the dust Of unlived dreams, where passion turned to ash, And purpose lost its vital, vibrant dash.

The Drifting Boat No anchor dropped, no sail unfurled with might, Just currents caught, by day and through the night. From fleeting whim to transient, dull pursuit, No roots struck deep, no lasting, patient fruit. The job half-learned, the friendship left to fade, A thousand starts, no masterpiece was made. They watched the world from easy, shifting sands, A life unshaped by purposeful commands. The days dissolved like mist upon the pane, A gentle glide, yet ending still in plain And quiet sorrow, for the "might have been," A life untouched, uncolored, and unseen.

The Shadowed Room Within four walls, a spirit chose to hide, From fear of judgment, or what lurked outside. The brilliant mind, that never shared its thought, The loving heart, no comfort ever brought. Perfection's chill, a constant, frozen hand, Prevented action, stifled every plan. They watched the brave ones stumble, fall, and rise, While safe within their self-imposed disguise. No risks were taken, lessons left unlearned, A timid flame, that never brightly burned. The world passed by, a vibrant, distant hum, And solitude became their final sum.

The Endless Chase A thirst unquenched, for more, and then for more, They knocked on every gilded, hollow door. For status gained, for power's fleeting kiss, For constant pleasure, purchased, never bliss. They climbed the peak, then sought another height, Ignoring sunsets, for the urban light. The human touch, the quiet, soulful need, Were trampled down by avarice and speed. They owned the world, yet felt an inner void, A grand possession, ultimately destroyed By its own nature, hungry and unkind, Leaving just echoes of an empty mind.

These paths converge in autumn's fading light, A silent knowing, in the lonely night. The bitter taste of chances left behind, A life unchosen, for the soul to find. For what is life, if not a vibrant quest, To breathe, to strive, to love, to truly test The bounds of self, beyond the easy sway, And cast a shadow, on a brighter day?

Gemini Apps

chat

The Ledger of a Greedy Heart

The ledger's open, pages worn and thin, A life laid bare, where endless wants pulled in. No grand designs of purpose, deep and true, Just constant hunger, chasing something new. The early years, a glint of nascent might, Soon dimmed beneath desire's blinding light.

I built my world with grasping, eager hands, Acquiring fortunes, plots of fertile lands. Each fleeting whim, a mandate to pursue, Another luxury, another view. The finest fabrics, meals of rich delight, Consumed until the day blurred into night. And still, a gnawing, restless, empty space, Unfilled by all the treasures I'd embrace.

I saw the struggling, felt their muted plea, But turned my gaze, consumed by only me. The planet groaned, its bounty plundered fast, My hurried footprint, far too deeply cast. The gentle whispers of a soul in need, Were drowned beneath the thrum of selfish greed. I chased the "more," while "enough" slipped away, Lost in the clamor of a grasping day.

Now silence falls, the fevered chase is done, And ghosts of chances whisper, one by one. The friendships thin, neglected for the gain, The quiet moments, sacrificed to pain. No deep contentment, only hollow pride, For all the riches held and kept inside. The sum total: not joy, but bitter sting, The barren harvest that a greedy life can bring.

The Echoes of a Consumed Life

The clock ticks back, each chime a hollow plea, For moments spent on what would simply be A fleeting joy, a plastic, polished gleam, The substance lost within a waking dream. My ledger etched, not with the deeds I'd done, But tally marks of battles dearly won To own the next, the shinier, the new, While silent, life's true treasures slipped right through.

The closets groan, a monument to want, Each folded shirt, a silent, mocking taunt. The shelves sag low with gadgets obsolete, Their fleeting promise, now a cold defeat. I chased the ads, the whispers of desire, Adding fuel to an insatiable fire. "This brings content," the glossy pages cried, But deep inside, a gnawing void resided.

My memories now feel paper-thin and vast, Of purchases, no true experience cast. The landscapes blurred, seen through a screen's soft haze, The conversations lost in retail's maze. Did I embrace a sunset, truly feel the rain? Or was my mind just planning what to gain? The laughter shared, the quiet, tender touch, Were these exchanged for owning far too much?

A bitter tang, this autumn of my years, To count the cost, drowned in unbidden tears. The time, the earth, the spirit I allowed To fade beneath consumption's heavy shroud. No legacy of purpose, deep and true, Just mountains built of things I journeyed through. Oh, for a breath, a chance to turn the tide, And fill this lingering emptiness inside, With moments lived, not merely bought and sold, A story rich, not just of silver, but of gold.

The table groans, a king's array, Of plenty piled, for yesterday. The feast was grand, the moments bright, Lost in the glow of appetite. Each dish devoured, a sweet excess, A fleeting thrill, a rich caress.

But now the dawn, a cruel, stark light, Reveals the errors of the night. My stomach churns, a rebel's drum, Against the bounty overcome. A heavy head, a fevered skin, The toxic wages from within.

The closet bursts, a vibrant hue, Of garments bought, both old and new. The screens once glowed with endless shows, A thousand plots, where pleasure flows. The gadgets gleam, a silent plea, To captivate, and set me free.

But freedom's bought at such a cost, In energy and moments lost. A restless mind, a weary soul, Consumed by all, yet far from whole. The fleeting joy, a shallow stream, Compared to life's profounder dream.

This bitter taste, this aching dread, The phantom weight of things unsaid, And things unlived, for things acquired, A hollow throne, a self-made pyre. The lesson learned, though hard and deep, Of seeds I've sown, and now must reap.

For sustenance is not mere greed, And true content, a simpler creed. To seek what serves, and truly fills, Beyond the market's gilded thrills. This sickness whispers, clear and grim: The deepest joys are found within.

The Feast of Emptiness

They fill their plates with shimmering vice, Greed's bitter taste devours the night, Their souls lie hungering for more.

The bottle calls in hollow tones, Liquor floods each trembling vein, They drown regret to numb the pain.

Bright screens glare with lustrous lure, Distraction's web ensnares the mind, Each swipe a promise left behind.

Coins clatter on scarred felt, Chasing hope through empty dreams, Gambling hearts with tethered strings.

Pills rattle in plastic shells, Ecstasy bites with hollow grin, Addiction's grasp begins within.

They gorge on sweetness, never sated, Gluttony swells each aching chest, Comfort's feast deprives its guest.

Mirror's smile becomes a mask, Narcissism blooms in glassy eyes, Self-love turned to bitter lies.

A world built on heated flesh, Hedonism's carnival rolls on, Pleasure chained, the heart is gone.

Immediate thrills demand their pound, Now becomes the only creed, Tomorrow's grave they scarcely heed.

They place themselves before all else, Selfishness cloaked in polite dress, No other need can theirs suppress.

Godless prayers are cast to stone, Belief dissolved in lust and greed, Faith abandoned in the seed.

Lines drawn fine with powdered white, The devil's lullaby in veins, Euphoria with tethered chains.

Drunken laughter, staggering steps, Careless souls in midnight's haze, Dawn reveals regret's sharp gaze.

They scratch their skin, they scar their minds, Self-abuse: a silent scream, A broken hope in shattered dreams.

Cigarette smoke curls with each breath, Poison whispers at the lip, A slow descent they cannot skip.

They chase the next electric jolt, Artificial dopamine, Lonely hearts bereft of home.

Mountains of plastic, waste untold, Consumption's monument to greed, Tomorrow buried by their need.

They worship sale signs, flashing red, Material gods bow at their feet, A fleeting rush they cannot keep.

Anger flares like wildfire's tongue, They scorch the ones they claim to love, Hearts charred by wrath from above.

Comfort comes on padded thrones, They numb the world with lavish ease, Empathy falls to its knees.

They hoard their hours in empty rooms, Countless toys to fill the void, A fleeting joy soon destroyed.

Perfume masks the stench within, Vanity's fine fragrant dye, Concealing cracks that multiply.

They post their victories for the world, Chasing likes in shallow fame, Approval's fire to feed the flame.

Envy prowls behind each scroll, Coveting another's smile, A poison plucked from bitter guile.

They pit their worth on pointed scales, Comparisons like jagged blades, Self-esteem in debt that never fades.

In whispered bets and secret trades, Their fortunes rise on strangers' fall, A heartless game where none stand tall.

They squander hours in silver fog, Smoking weed to blur the lines, Time dissolves in haze and wine.

They live for nights that never end, Club lights pulse with vacant cheer, Music hides the silent fear.

They purchase plastic paradise, Exotic dreams in glossy wraps, Only to find regret's relapse.

They swallow pills to kill the ache, Antidepressant lullabies, Yet sadness thrives in painted skies.

They hoard the latest trending thing, Then toss it when desire fades, A landfill grows from shallow trades.

Anger stokes the furnace deep, Road-rage horns and shouted spite, Humanity lost in spiteful night.

They binge on shows until they choke, Banish boredom with each new scene, Life reduced to glowing screen.

They pierce their flesh with careless hands, Tattoos of sorrow, scars of shame, Each line a testament to blame.

They pray to idols made of gold, Brands that gleam with empty shine, Merchandise becomes divine.

They chase the perfect sculpted form, Gym sweat baptized in vanity, A temple built on vanity.

They barter sleep for restless thought, Midnight scrolling kills repose, Morn's regret in heavy throes.

They taste the spice of forbidden touch, Promiscuous nights, transient heat, Affection sold on restless streets.

They fill their lungs with smoky haze, Vaping clouds of flavored lies, Youth surrendered to disguise.

They drown their guilt in candy bar, Sugar highs to mask the ache, A child's solace quickly breaks.

They drown in streams of data sprawl, Newsfeeds feed their darkest fears, Chaos pumped through wired ears.

They worship youth, they fear the age, Anti–time's relentless march, Wrinkles met with plastic arch.

They numb their hearts with endless toil, Work's applause becomes their breath, Burnout stalks just life and death.

They gamble futures on quick schemes, Ponzi dreams in frantic pitch, Hopes washed down in fraudulent ditch.

They lock their doors on neighbor's cries, Comfort barricaded tight, Human kindness lost to fright.

They drown in Netflix's velvet grip, Marathon of borrowed lives, Own story barely survives.

They wrap their wounds in latex masks, Cosmetic cuts that never heal, Inner pain they cannot feel.

They chase the flash of fleeting wealth, Crypto castles built on sand, Dreams evaporate at hand.

They brandish phones like sacred scrolls, Notifications drum their pulse, Attention sold for impulse.

They hoard the cheers of shallow crowds, Adulation fools the mind, True worth left too far behind.

A pill to wake, a pill to sleep, Chemical crutches day and night, Freedom sold in measured might.

They block the world with noise machines, Earbuds drown the honest voice, Silence lost by painful choice.

They pawn their youth for flashy cars, Leather seats, a tyrant's throne, Debt-bound journeys they roam alone.

They spew their hate on anonymous screens, Trolls with venom in their veins, Human decency drained in chains.

They carve their names in transient sand, Online legacies so frail, Waves erase each digital trail.

They flavor tears with bitter wine, Mournful sips of sorrow's print, Hope's faint echo fades to lint.

They fast for fame, they fast for likes, Sacrificing truth and rest, The price of influence confessed.

They hoist the flag of righteous ire, Anger bleeds on every side, No bridge can span the widening tide.

They stitch their worth in branded cloth, Logo stitched in price and pride, Value measured far and wide.

They swallow lies from tailored ads, False promises in pretty hues, A world manipulated for its dues.

They haunt the aisles of midnight malls, Sale signs beckon hungry eyes, Wallets empty under lies.

They code the world in binary dreams, AI whispers in each ear, A future guided by its fear.

They sell their youth to tanning beds, Bronze illusions bake their flesh, Cancer's seed they softly mesh.

They drown regret in midnight rides, Speed becomes their fleeting song, Trails of smoke where morals throng.

They sharpen words like sharpened blades, Argument's a battlefield, Empathy becomes concealed.

They sculpt their thoughts in echo rooms, Repetition numbs the mind, Originality left behind.

They feed resentment's hungry maw, Grudges second every beat, Forgiveness banished in retreat.

They practice hate as holy creed, Preach division at the door, Compassion leaves the floor.

They shuffle pills in blister packs, Medication for each mood, Authentic self misconstrued.

They chase the sun in jet-set life, Passport stamped with restless feet, Home becomes a foreign street.

They carve their worth in peers' applause, Every smile a currency, Self-esteem bound to decree.

They feast on memes, their humor cold, Laughter canned in packet jokes, Human warmth dissolves in pokes.

They cradle screens in tender hands, Digital cradles rock their dreams, Real touch lost in data streams.

They brandish credit as a sword, Debt their shield in ravenous quests, Treasures lost in interest.

They strain to hear a pulpit's call, Spiritual void in empty pews, Faith replaced by corporate views.

They chase the high of viral fame, Vlog and vlog for fleeting fame, Identity lost in name.

They pawn their moral compass quick, Ethics bent to suit the bid, A conscience sold, a soul unhid.

They drown the past in tequila dreams, Blackout nights with scarlet mind, Regret's ghost lingers behind.

They chase the drag of foreign tongues, Cultures swallowed by prestige, Diversity fades like mist.

They cultivate their sculpted bones, Anorexic gods of glass, True beauty lost in hollow mass.

They whisper lies to tranquilize, Self-deception's lullaby, Truth slips softly from the sky.

They ransom sleep for midnight oil, Burning vigor to the bone, Rest abandoned, overthrown.

They pack their schedule tight with noise, No silence left to simply be, Solitude they fail to see.

They sharpen phones with pointed glance, Eyes glazed in artificial shine, The world outside begins to pine.

They traffic gossip at the gate, Rumors feed their restless tongues, Integrity left unsung.

They cower when the storm draws near, Anxiety their closest friend, Peace a stranger to the end.

They strain for more when plenty waits, Gratitude a fleeting guest, Contentment kept in shallow chest.

They bury love in tangled sheets, Intimacy turned hollow feat, Hearts unattended in retreat.

They prop up youth with silent lies, Anti–age in every bottle, Aging met by empty throttle.

They shred the past in vanity's shred, Polished scars on curated skin, Memory lost in trending spin.

They cling to titles, power, rank, Authority their anchor cold, Humanity bought and sold.

They stuff their lungs with toxic fumes, Factory skies their daily breath, Nature choked in silent death.

They chase the sun but fear the warmth, Melting walls of restless doubt, Isolation all about.

They barter secrets for a glance, Privacy their dying art, Public hearts remain apart.

They drown desire in plastic seas, Micro waste in every tide, Ocean's tears they spam and hide.

They prize excuses over truth, Accountability long lost, Responsibility the cost.

They cheer distraction's every call, Busy hands that never rest, Quiet moments fail the test.

They trade their morals for success, Paved with bodies in the race, Conscience buried without trace.

They hoard regrets in silent vaults, Memories locked in iron bars, Hope extinguished by their scars.

And in their feast of emptiness, They dine on sins they cannot flee, A hollow world of misery.

chat

Consumption, especially overconsumption or unsustainable consumption, has a wide range of negative effects across environmental, social, economic, and individual health domains. Here are some key examples:1. Environmental Degradation:Resource Depletion: The extraction of raw materials(minerals, metals, fossil fuels, timber) to produce goods is happening at a rate faster than the Earth can replenish them. This leads to the depletion of finite resources.

Pollution:

Air Pollution: Manufacturing processes, transportation (shipping, planes, trucks), and energy production for consumption release greenhouse gases and other pollutants into the atmosphere, contributing to climate change, smog, and respiratory illnesses.

Water Pollution: Industrial waste, agricultural runoff (driven by food consumption), and improper disposal of consumer products contaminate water sources, harming ecosystems and human health. The clothing industry alone is a significant contributor to water pollution.

Land Pollution/Waste Crisis: Excessive consumption generates massive amounts of waste, leading to overflowing landfills and plastic-clogged oceans. Much of this waste, especially plastics and electronics (e-waste), is non-biodegradable and can leach toxic chemicals into the environment.

Biodiversity Loss: Demand for land, raw materials, and agricultural expansion (to feed growing populations and consumption patterns) leads to deforestation, habitat destruction, and the extinction of species.

Climate Change: The entire lifecycle of products, from extraction to disposal, contributes to greenhouse gas emissions, accelerating global warming and its associated impacts like extreme weather events, sea-level rise, and food insecurity.

2. Social Consequences:

Inequality: Wealthier nations and individuals consume significantly more resources than poorer ones, exacerbating global inequalities. This can lead to the exploitation of resources and labor in developing countries.

Poor Labor Standards: The drive for cheaper goods often leads companies to outsource production to regions with lax labor laws, resulting in low wages, long hours, and dangerous working conditions for workers.

Social Isolation/Mental Health: While consumption can provide temporary satisfaction, a constant preoccupation with acquiring more can lead to increased anxiety, stress, and a focus on material possessions over meaningful relationships or experiences. It can also contribute to feelings of inadequacy or keeping up with others.

Community Strain: In cases of addiction (e.g., alcohol or drug consumption), there are severe social consequences like strained relationships, increased risk of domestic violence, financial problems, and a burden on community resources like healthcare and law enforcement.

3. Economic Drawbacks:

Planned Obsolescence: Many products are intentionally designed to have a short lifespan, forcing consumers to constantly replace them. This drives continuous consumption and generates more waste.

Increased Debt: Consumers may take on significant debt to finance their consumption habits, leading to financial instability for individuals and households.

Economic Instability: While consumer spending is a key driver of economic growth, unsustainable consumption patterns can create bubbles and lead to resource scarcity, potentially disrupting economic stability in the long term.

Misallocation of Resources: Resources that could be used for essential services or sustainable development may be diverted to produce non-essential consumer goods.

4. Health Impacts (Individual and Public Health):

Direct Exposure to Pollutants: Air and water pollution from consumption-related activities can directly cause respiratory illnesses, cancers, neurological disorders, and other health problems.

Food-Related Health Issues: Overconsumption of certain foods, especially processed and unhealthy options, contributes to rising rates of obesity, heart disease, diabetes, and other diet-related illnesses. The industrial production of food also has environmental and health consequences.

Indirect Health Impacts of Climate Change: As consumption drives climate change, the resulting impacts (e.g., extreme heat, increased infectious diseases, food and water scarcity) directly affect human health and well-being.

Mental Health: The stress of consumerism, financial strain, and the impacts of environmental degradation can all contribute to mental health issues like anxiety and depression.

These negative effects are often interconnected, creating a complex web of challenges that highlight the urgent need for more sustainable consumption patterns.

Why this activity?

This activity was saved to your Google Account because your Gemini Apps Activity setting was on while using Gemini Apps.

info_filled

If you use a shared device or sign in with multiple accounts, activity from another account may appear here. Learn more

Prompted specific examples of my own over consumption in my life of 8 decades

Details

event

July 11 at 8:00 PM

apps

Gemini Apps

chat

Please provide me with some information about your consumption habits over the past eight decades. Since I don't have access to your personal information or historical data, I cannot give you specific examples of your overconsumption.

However, I can give you common examples of overconsumption that many people in developed countries, particularly over several decades, might experience. As you read these, you can reflect on whether they apply to your own life:

Common Areas of Potential Overconsumption Over 8 Decades:

Vehicles:

Owning multiple cars simultaneously: Did you often have more vehicles than strictly necessary for your household's needs?

Frequent car replacement: Upgrading to new cars every few years even when the old one was perfectly functional.

Choosing larger, less fuel-efficient vehicles: Opting for SUVs or trucks when a smaller, more efficient car would suffice for most daily needs.

Driving for short distances: Using a car for trips that could have been walked, biked, or covered by public transport.

Clothing and Fashion:

"Fast Fashion": Purchasing trendy, inexpensive clothes that go out of style quickly or wear out after a few washes, leading to frequent replacement.

Excessive wardrobe size: Owning far more clothes, shoes, and accessories than you regularly wear.

Buying for "just in case" or "one-off" events: Acquiring outfits for specific occasions that are rarely worn again.

Electronics and Gadgets:

Frequent upgrades: Replacing phones, computers, televisions, or other electronics every few years when older models were still functional.

Accumulation of devices: Having multiple televisions, tablets, or gaming consoles that are not all regularly used.

Buying the "latest and greatest": Feeling compelled to acquire new models even when the improvements are marginal.

Home Goods and Decor:

Excessive furniture: Owning more furniture than truly needed for comfortable living, perhaps filling larger homes or storage units.

Decorative items: Continuously buying knick-knacks, art, or seasonal decorations that accumulate over time.

Kitchen gadgets: Purchasing specialized kitchen appliances that are used infrequently.

Storage unit reliance: Needing to rent external storage space because your home can't contain all your possessions.

Food and Groceries:

Food waste: Regularly buying more food than you or your household can consume before it spoils.

Over-sized portions: Consistently preparing or ordering larger portions than necessary, leading to leftovers that may or may not be eaten.

Frequent dining out/takeout: Consuming many meals prepared outside the home, which often has a larger environmental footprint and can lead to less mindful consumption.

Excessive packaging: Consistently choosing products with single-use or excessive packaging.

Leisure and Hobbies:

Accumulation of hobby equipment: Buying multiple sets of golf clubs, fishing gear, or crafting supplies, or new models when older ones are still perfectly usable.

Disposable leisure items: Frequently purchasing items for one-time use during leisure activities (e.g., disposable picnic supplies, single-use camping gear).

Energy and Utilities:

Maintaining larger homes: Living in homes that are significantly larger than your actual needs, requiring more energy for heating and cooling.

Leaving lights/electronics on: Not consistently turning off lights, televisions, or other electronics when not in use.

High thermostat settings: Keeping heating or air conditioning at extreme settings, leading to higher energy consumption.

Long showers/baths: Using excessive amounts of water for personal hygiene.

Travel:

Frequent air travel for leisure: Taking numerous flights for vacations, especially to distant locations, accumulating significant carbon footprints.

Luxury travel accommodations: Consistently choosing high-resource-intensive travel options (e.g., large cruise ships, all-inclusive resorts with high waste generation).

To identify your own specific examples, I encourage you to reflect on your past purchasing decisions, your current possessions, and your daily habits. Think about:

"Do I really need this?" before buying.

"How often do I actually use this?" for items you own.

"Could I have borrowed this or repaired the old one?"

"What happens to this item when I'm done with it?"

By considering these questions, you can pinpoint areas where you might have consumed more than was truly necessary or sustainable over your lifetime.

Why this activity?

This activity was saved to your Google Account because your Gemini Apps Activity setting was on while using Gemini Apps.

info_filled

If you use a shared device or sign in with multiple accounts, activity from another account may appear here. Learn more

Prompted examples of gluttony

Details

event

July 11 at 8:02 PM

apps

Gemini Apps

chat

Gluttony, traditionally one of the "seven deadly sins," refers to excessive indulgence and overconsumption, particularly of food and drink, to the point of waste or harm. While often associated with eating, it can extend to any form of excessive desire or consumption of material pleasures.

Here are examples of gluttony:

- Food and Drink (Most Common Association):

Eating far beyond satiation: Continuing to eat a large meal even when already full and uncomfortable, simply because the food is delicious or available.

Mindless eating/binging: Consuming large quantities of snacks, fast food, or unhealthy items without true hunger, often due to boredom, stress, or emotional reasons, leading to feeling unwell afterwards.

Obsessive pursuit of luxurious or rare foods: Constantly seeking out the most expensive, exotic, or elaborately prepared dishes, not for nourishment, but for the sheer indulgence and display, potentially at the expense of others or sustainability.

Excessive drinking: Consuming alcoholic beverages to the point of intoxication or beyond, leading to impaired judgment, health problems, and potential harm to oneself or others (drunkenness is often considered a form of gluttony).

Food waste due to over-preparation: Cooking or ordering far more food than can reasonably be consumed, leading to significant leftovers that are ultimately thrown away.

Eating too quickly or greedily: Devouring food without savoring it, often leading to discomfort and still wanting more due to the speed of consumption.

The "all-you-can-eat" buffet mentality: Approaching such buffets with the goal of eating as much as possible, regardless of hunger, simply because it's available and paid for.

Beyond Food (Broader Interpretation):

Material Gluttony (Greed/Excessive Acquisition):

"Shopaholic" behavior: Compulsively buying clothes, gadgets, or other consumer goods far beyond what is needed or can be used, leading to excessive clutter, debt, and dissatisfaction.

Hoarding: Accumulating an excessive amount of possessions to the point where it clutters living spaces and interferes with daily life.

Collecting to an extreme: While collecting can be a hobby, it becomes gluttonous when the acquisition of items takes over one's life, consumes excessive financial resources, and becomes an insatiable desire without true enjoyment.

Entertainment/Digital Gluttony:

Binge-watching to excess: Spending entire days or nights consuming hours upon hours of television shows, movies, or video games, neglecting responsibilities, sleep, or social interaction.

Compulsive internet/social media use: Spending an inordinate amount of time scrolling, Browse, or engaging with online content, even when it leads to feelings of anxiety, comparison, or isolation.

Power/Dominion Gluttony:

Ruthless ambition: An insatiable desire for power, control, or recognition, leading to unethical behavior, exploitation of others, and a constant need to accumulate more influence.

Historical and Literary Examples:

Roman Feasts (Historical): Accounts of ancient Roman feasts often describe elaborate banquets where participants would gorge themselves, sometimes inducing vomiting to make room for more food.

Augustus Gloop (Charlie and the Chocolate Factory): This fictional character is a classic portrayal of gluttony, falling into a chocolate river due to his insatiable desire to consume it.

King Henry VIII: Known for his massive appetite and large stature, his diet was often cited as an example of royal gluttony, consuming vast quantities of meat and ale.

Esau selling his birthright (Biblical): In the Bible, Esau sells his valuable birthright for a single bowl of lentil stew, illustrating prioritizing immediate, fleeting gratification over long-term wisdom and spiritual value.

In essence, gluttony is about a lack of self-control and an unhealthy attachment to pleasure or material things, consuming beyond what is necessary, reasonable, or beneficial, often with negative consequences for the individual, others, or the environment.

Why this activity?

This activity was saved to your Google Account because your Gemini Apps Activity setting was on while using Gemini Apps.

info_filled

If you use a shared device or sign in with multiple accounts, activity from another account may appear here. Learn more

Prompted specific examples of the types of things that lead to wasted lives

Details

event

July 11 at 8:03 PM

apps

Gemini Apps

chat

The idea of a "wasted life" is deeply personal and often rooted in regret, unfulfilled potential, and a sense that time and opportunities were not used meaningfully. While success is often measured by external achievements, a "wasted life" often refers to an internal feeling of dissatisfaction and lack of purpose.

Here are specific examples of types of things that can lead to wasted lives, categorized for clarity:

I. Lack of Self-Awareness and Authenticity:

Living for others' expectations: Constantly trying to please parents, partners, society, or peers, rather than discovering and pursuing one's own desires and values. This can lead to choosing careers, lifestyles, or even relationships that don't genuinely align with who you are. (e.g., Becoming a lawyer because your parents wanted you to, even though your passion was art).

Ignoring one's true passions and talents: Suppressing creative urges, intellectual curiosities, or unique skills because they aren't seen as "practical" or lucrative, leading to a feeling of unfulfilled potential.

Lack of self-reflection: Never taking the time to understand what truly brings joy, meaning, and purpose, instead drifting through life on autopilot.

Fear of being oneself: Hiding aspects of one's personality, beliefs, or identity due to fear of judgment, rejection, or not fitting in.

II. Habits and Mindsets that Hinder Growth:

Procrastination: Consistently delaying important tasks, dreams, and goals, leading to missed opportunities and a lingering sense of unaccomplishment. (e.g., Always planning to write that book, learn that language, or start that business, but never actually starting).

Excessive distraction and instant gratification: Spending an inordinate amount of time on social media, binge-watching TV, playing video games, or engaging in other non-productive activities that provide fleeting pleasure but no lasting fulfillment.

Perfectionism and fear of failure: Being so afraid of making mistakes or not achieving perfect results that one never takes risks or even begins important endeavors.

Constant comparison to others: Measuring one's life against the seemingly perfect lives presented on social media or by societal standards, leading to feelings of inadequacy, envy, and dissatisfaction.

Lack of discipline and consistency: Starting many things but finishing few, leading to a scattered life and a lack of mastery in any area.

Negative self-talk and limiting beliefs: Believing one isn't capable, smart enough, or worthy of success and happiness, which becomes a self-fulfilling prophecy.

Holding grudges and resentment: Carrying emotional baggage from the past, which drains energy and prevents moving forward with joy and peace.

III. Relationship and Social Factors:

Staying in toxic relationships: Remaining in friendships, romantic relationships, or even family dynamics that are abusive, draining, or stifle personal growth.

Neglecting meaningful relationships: Prioritizing work, money, or distractions over investing time and effort in deep, authentic connections with loved ones and friends. (A very common regret in old age).

Isolation and lack of community: Lacking a sense of belonging or connection to something larger than oneself, leading to loneliness and a feeling of meaninglessness.

IV. Career and Financial Pitfalls:

Being stuck in a dead-end job: Staying in unfulfilling work for security or comfort, even if it drains energy and provides no sense of purpose or growth.

Prioritizing money over meaning: Chasing wealth or status without considering if the path aligns with personal values or brings genuine satisfaction. (e.g., Working grueling hours in a high-paying job you hate, at the expense of health, relationships, or personal time).

Poor financial management: Accumulating debt, living beyond one's means, or failing to plan for the future, leading to chronic stress and limited choices.

V. Health and Well-being Neglect:

Ignoring physical health: Neglecting exercise, healthy eating, and sufficient sleep, leading to chronic illness, low energy, and a diminished quality of life.

Neglecting mental and emotional health: Avoiding addressing issues like anxiety, depression, or trauma, which can severely impair one's ability to engage with life fully.

Substance abuse or addiction: Relying on drugs, alcohol, or other addictive behaviors to cope with life's challenges, leading to a downward spiral and loss of control.

Ultimately, a "wasted life" often stems from a combination of these factors, where individuals lose agency over their choices, become victims of circumstance or their own habits, and fail to cultivate a life that reflects their unique potential and brings them authentic joy and purpose.

Why this activity?

This activity was saved to your Google Account because your Gemini Apps Activity setting was on while using Gemini Apps.

info_filled

If you use a shared device or sign in with multiple accounts, activity from another account may appear here. Learn more

Prompted the hollowness of being just a consumer

Details

event

July 11 at 8:04 PM

apps

Gemini Apps

chat

The phrase "the hollowness of being just a consumer" captures a profound critique of modern consumer culture and its impact on individual well-being and societal values. It refers to the idea that solely defining oneself by what one buys and consumes ultimately leads to a superficial, unfulfilling, and ultimately empty existence.

Here's a breakdown of the "hollowness" and why it occurs:

1. The Hedonic Treadmill (or Hedonic Adaptation):

The initial high fades: The thrill of a new purchase (a new car, phone, piece of clothing, etc.) is real but remarkably short-lived. We experience a surge of happiness or satisfaction, but quickly adapt to the new item. What was once exciting becomes the new normal.

Constant craving: Because the initial "high" fades, we are compelled to seek the next new thing, creating an endless cycle of desire and acquisition. It's like running on a treadmill – you exert a lot of effort, but you don't actually move forward in terms of lasting happiness.

Never enough: The problem isn't the item itself, but the reliance on external objects for internal fulfillment. Material possessions, by their nature, cannot provide enduring happiness or meaning.

2. Superficial Identity and Self-Worth:

"You are what you buy": Consumer culture often promotes the idea that our identity and worth are tied to the brands we wear, the cars we drive, the homes we live in, and the experiences we post online. This creates a fragile sense of self, constantly dependent on external validation.

Comparison culture: Social media amplifies this by showcasing idealized versions of others' lives, filled with consumer goods and experiences. This fuels constant comparison, leading to feelings of inadequacy, envy, and a relentless pressure to "keep up with the Joneses."

Lack of intrinsic value: When self-worth is based on material possessions, it detracts from developing intrinsic qualities like kindness, empathy, creativity, resilience, or intellectual curiosity. These are deeper, more stable sources of self-esteem.

3. Distraction from Deeper Issues:

Escapism: Shopping and acquiring can become a coping mechanism, a form of "retail therapy" used to temporarily numb feelings of boredom, anxiety, loneliness, or dissatisfaction with life. However, it doesn't address the root causes of these feelings, only postpones them.

Avoidance of self-reflection: The constant pursuit of external stimulation through consumption can prevent individuals from engaging in meaningful self-reflection, understanding their true needs, or facing uncomfortable truths about their lives.

4. Erosion of Meaningful Connections:

Prioritizing things over people: A focus on material acquisition can lead to neglecting relationships, experiences, and community engagement. Time and energy are often diverted from building deep connections to earning money for more consumption.

Shallow interactions: Conversations might revolve around purchases or what others own, rather than genuine connection, shared experiences, or intellectual discourse.

5. Financial Stress and Debt:

The cost of the chase: Maintaining a consumerist lifestyle often requires constant earning and can lead to significant financial stress, debt, and a feeling of being trapped by one's expenses. This directly contradicts the promise of happiness through acquisition.

Work-life imbalance: The need to fund a consumerist lifestyle can drive people to overwork, sacrificing leisure, health, and family time.

6. Environmental Guilt and Anxiety:

Awareness of impact: As awareness of climate change and environmental degradation grows, individuals who consume excessively may experience "eco-anxiety" or guilt over their contribution to these problems, further contributing to the hollowness.

The Counterpoint: Beyond Consumption

The hollowness of being just a consumer highlights the importance of shifting focus from "having" to "being" and "doing." Fulfillment often comes from:

Experiences: Travel, learning new skills, spending time in nature, engaging in hobbies.

Relationships: Nurturing deep, authentic connections with family and friends.

Purpose and Meaning: Engaging in work or activities that align with one's values and contribute to something larger than oneself.

Personal Growth: Continuous learning, self-reflection, and developing intrinsic qualities.

Creativity and Contribution: Producing, building, or giving back to the world.

In essence, being "just a consumer" implies a passive role, where one is primarily on the receiving end of commercial messages and products. A truly rich and fulfilling life, however, requires active engagement, creation, connection, and a sense of purpose that transcends mere acquisition.

long poem of at least 100 stanzas about all the things people do to sicken themselves with consumption and waste in lives that have little meaning but for greed, anger, gluttony, narcissism, hedonism, and immediate gratification, as well as selfishness and Godlessness drug use, drunkenness, carelessness, self abuse. about all the things people do to sicken themselves with consumption and waste in lives that have little meaning but for greed, comfort, immediate gratification, and hedonism, gambling, addiction, distraction

one poem of at least 100 stanzas about all the things people do to sicken themselves with consumption and waste in lives that have little meaning but for greed, anger, gluttony, narcissism, hedonism, and immediate gratification, as well as selfishness and Godlessness drug use, drunkenness, carelessness, self abuse. about all the things people do to sicken themselves with consumption and waste in lives that have little meaning but for greed, comfort, immediate gratification, and hedonism, gambling, addiction, distraction

Absolutely—I can craft a long-form poem capturing those themes with vivid imagery and a lyrical edge. Here's the beginning of a 100-stanza epic poem titled "The Feast of Emptiness". I'll share the first portion now, then we can continue with more if you like the tone and style:

The Feast of Emptiness (Stanzas 1–20)

1 They gather round the golden tray, Where hunger meets deceitful prey. A banquet set with borrowed grace, Their mirrored masks begin to face.

2 The wine they sip, a poison sweet, To drown their doubts in soft defeat. Each drop a whispered lullaby, That sings of pleasures gone awry.

3 The dealers smile, the gamblers nod, As coins and time are blindly trod. Addiction's hands stretch long and thin, A velvet noose they slip within.

4 They build their lives on fleeting thrills, Fast cars, bright lights, and loaded pills. But still the silence echoes loud, Behind the music, under crowds.

5 Comfort becomes a padded cell, Where thought and conscience cease to dwell. Luxury blinds with silken thread, As empathy lies cold and dead.

6 They dance in clubs with vacant eyes, Where false euphoria multiplies. The beat is nothing but escape, A breathless gasp they try to shape.

7 Godless rituals, forged in haste, Sacrifices made to taste. The idols now wear human skin, And preach of pleasure, rage, and sin.

8 Gluttony in gilded halls, Bloating souls behind the walls. Feasting on what cannot feed, Devouring dreams for shallow need.

9 Screens glow like unholy flame, Distraction wears a thousand name. They scroll until their hearts go numb, And wonder where the joy came from.

10 Drunken nights, the hallway crawl, Truth erased by alcohol. A morning filled with heavy shame, No one to blame, no one to name.

11 They spend, consume, then crave anew, The cycle spins its bitter hue. In shopping carts and checkout lines, They chase what never realigns.

12 Self-abuse and silent screams, The death of self in shattered dreams. Each mirror cracked, each razor kissed, In search of warmth that won't exist.

13 Comfort crumbles into dust, Once built on greed, desire, and lust. They clutch their things but feel so lost, And never count the mounting cost.

14 Plastic smiles and hollow praise, Echo through the modern maze. Influence bought with every view, But purpose buried under new.

15 Gambling hearts and gambling minds, They risk the soul for hollow finds. Chips fall where their morals lay, And innocence is swept away.

16 With every drug, they dim their light, And welcome in eternal night. They numb the ache they've never named, Unknowing they are self-enslaved.

17 Greed, the god they bow before, Whispers always "More, more, more." No offering ever satisfies, No shrine can cleanse their bloodshot eyes.

18 Anger flares like sudden flame, Another soul for them to maim. The blame game plays, the fists will fly, And all the truth they once knew dies.

19 Hedonism paints the town, With smiles that drip and faces drown. The thrills are short, the crashes hard, And no one keeps their soul unmarred.

20 So onward still the march of waste, Of life devoured with fevered haste. They chase a dream that never stays— A ghost that fades in morning haze.

Let me know if you'd like more—there's plenty of room left in the descent, and perhaps a glimmer of redemption. Or we can turn this into a fully bound cycle, a story told from ruin to rebirth.

I've written a sestina titled "Something Sad About Serendipity" that follows the traditional form's complex structure. A sestina consists of six stanzas of six lines each, plus a three-line envoi (conclusion). The six end-words (see, chance, fate, doors, story, find) rotate in a specific pattern throughout the poem, creating a intricate weaving effect that mirrors the theme of serendipity itself.

The poem explores the melancholy paradox of serendipity—how the more we search for meaningful coincidences and connections, the more we might miss them when they actually occur. It touches on themes of missed opportunities, the timing of love and fate, and the bittersweet nature of looking back on the chances we didn't take or recognize.

i am an old man who consumed my life in all the falsities mentioned above. Write and autobiographical poem of at least 200 stanzas about consumption sickness. little meaning but for greed, anger, gluttony, narcissism, hedonism, and immediate gratification, as well as selfishness and Godlessness drug use, drunkenness, carelessness, self abuse. about all the things people do to sicken themselves with consumption and waste in lives that have little meaning but for greed, comfort, immediate gratification, and hedonism, gambling, addiction, distraction

Made in the USA
Monee, IL
04 November 2026

67385843R10203